JUSTICE, SOCIETY AND NATURE

D0143985

Environmental justice is about the fair distribution of good and bad environments to humans. Ecological justice is about fair distribution of environments among all the inhabitants of the planet. To speak of 'environmental' or 'ecological' justice means to recognise the value that an environment has for all creatures. An environment is comprised not only of people, but also non-human nature in all its abundance and diversity: animals and plants, landscapes and ecologies. An environment is not divisible like property but is fundamentally shared. Bad environments are dead, disintegrated, damaging to health. Good environments are alive, healthy and integrated.

Justice, Society and Nature examines the moral response which the world must now make to the ecological crisis if there is to be real change in the global society and economy to favour ecological integrity. From its base in the idea of the self, through principles of political justice to the justice of global institutions, the authors trace the layered structure of the philosophy of justice as it applies to environmental and ecological issues.

Moving straight to the heart of pressing international and national concerns, the authors explore the issues of environment and development, fair treatment of humans and non-humans, and the justice of the social and economic systems which affect the health and safety of the peoples of the world. Current grassroots concerns such as the environmental justice movement in the USA, and the ethics of the international regulation of development are examined in depth.

This book is essential for those seeking to understand current debates in environmental politics, justice and ecology. The authors take the debates beyond mere complaint about the injustice of the world economy, and suggest what should now be done to do justice to nature.

Nicholas Low is a Senior Lecturer in the Faculty of Architecture, Building and Planning at the University of Melbourne. **Brendan Gleeson** is a Research Fellow at the Australian National University, Urban Research Program.

JUSTICE, SOCIETY AND NATURE

An exploration of political ecology

Nicholas Low and Brendan Gleeson

London and New York

First published 1998
by Routledge
11 New Fetter Lane, London EC4P 4EE

Simultaneously published in the USA and Canada
by Routledge
29 West 35th Street, New York, NY 10001

Typeset in Garamond by Routledge
Printed and bound in Great Britain by
Biddles Ltd, Guildford and King's Lynn

British Library Cataloguing in Publication Data
A catalogue record for this book is available from the British Library

Library of Congress Cataloguing in Publication Data
A catalogue record for this book has been requested

ISBN 0–415–14516–3 (hbk)
ISBN 0–415–14517–1 (pbk)

In memory of Stuart Low,
a fellow traveller,
and for Ulrike

CONTENTS

CONTENTS

FIGURES

ACKNOWLEDGEMENTS

The authors wish to acknowledge the sources of inspiration, support and intellectual challenge for this book. Of course the book seeks to make a small contribution to a tradition of scholarship from which the authors draw much of their understanding of the issues and whose scope is indicated in the Bibliography. But there were many other helpers.

Tristan Palmer, then with Routledge, was the first to encourage the preparation of the book. Subsequent planning with Brendan Gleeson took place during a visit by Nicholas Low to the Department of Geography and Planning at the University of Otago at the invitation of Ali Memon, with whom the authors carried out some exciting preliminary brainstorming. Sarah Lloyd, our editor at Routledge, has always given encouragement and useful suggestions. Referees Ian Simmons and David Pepper provided most valuable advice on the structure and detail of the text in the later stages of the book.

Nicholas would like to acknowledge the support given by the Faculty of Architecture, Building and Planning at the University of Melbourne, especially its Dean, Ross King. The Faculty provided funds to travel to the Universities of Örebro and Umeå in Sweden where Nicholas spent time working on this book in fruitful discussion with Ingemar Elander at Örebro and Abdul Khakee at Umeå. Brendan Gleeson wishes to acknowledge the support given by the Geography Department of the University of Otago and Professor Ali Memon, also by the Urban Research Program at the Australian National University and Professor Pat Troy.

Thanks go also to our families who were always supportive and inspirational, especially, of course, our partners Elizabeth Low and Ulrike Gleeson. Nicholas would also like to thank his daughters Vinca Low, a student of both political ecology and classics, who found the epigraph for the book by the Roman poet Seneca, Jennifer Low for lively conversation and Ciannait Low, an environmentalist, for her love and support at the time of the tragic death of Nicholas's son Stuart.

My greatest pleasure was to scan the sky,
That noblest work of the great architect
Of infinite creation, Mother Nature,
Marking the motions of the universe,
The passage of the chariot of the sun,
The night's recurring phases, and the moon's
Bright orb encircled by the wandering stars,
The vast effulgence of the shining heavens.
Is all this glory doomed to age with time
And perish in blind chaos? Then must come
Once more upon the world a day of death,
When skies must fall and our unworthy race
Be blotted out, until a brighter dawn
Bring in a new and better generation
Like that which walked upon a younger world
When Saturn was the ruler of the sky.
That was the age when the most potent goddess,
Justice, sent down from heaven with Faith divine,
Governed the human race in gentleness.

SENECA

(From Octavia in *Four Tragedies*, translated by E.F. Watling,
Penguin Books, London, 1966, pp. 271–273 *Octavia*,
lines 385–404)

1

JUSTICE IN AND TO
THE ENVIRONMENT

Never allow yourself to be swept off your feet; when an impulse stirs, see first
that it will meet the claims of justice.

(Marcus Aurelius, *Meditations*)

INTRODUCTION

Justice is no mere abstraction. Finding justice and doing justice is a continuous
human task. It is the activity which in any society gives politics and the law
their purpose. The activity has both a material and a discursive dimension.
It has to do with what we are, what we do and what we say. What we are
and do is materially real. How we relate to others is discursively real, a
matter of communicated explanations via words. The struggle for justice is
about how we explain the basis of a good and proper relationship between
ourselves and others. In defining this relationship we define who and what
we are and who and what 'the other' is.

This task is continually expanding as new actors enter the struggle: new
perspectives are inscribed in the debates, new problems are defined for
society, in short, as politics in the broadest sense takes new forms. The question of justice is today being reshaped by the politics of the environment.
For the first time since the beginning of modern science we are having to
think morally about a relationship we had assumed was purely instrumental.
In the ancient world humans were seen as the instruments of nature or 'the
gods'. In the modern world the position was reversed; a disenchanted
'nature' became an infinite pool of resources to be made into things of use to
us. Today the relationship between humans and the rest of the natural world
is again being redefined. At the precise moment when it became clear that
we humans had this planet in the palm of our hand, it also became clear that
we are likewise held by it. Just when we became free and separated from the
earth we discovered the nature of our attachment to it.

The word 'environment' today seems scarcely adequate to describe that to
which we are attached. The planet, and indeed the universe, seems much
more than the surroundings of a human person to be defined only in relation

1

to the human. Yet our experience of nature is necessarily localised. Even from space a person may *view* the planet Earth as a whole but interact mostly with the little bit of nature she takes with her as 'life support'. We experience our own small part of 'nature' which we look out on, interact with, breathe, eat, drink, touch, hear and smell. This is our *environment*. It can be very good for us or very bad. Some people live and work in delightful environments. Others exist in oppressive and ugly environments. For some what was once a harmonious and healthy relationship with their environment is transformed suddenly into a risky and dangerous one.

Who are 'we'? There are two meanings of 'we': 'we the people' and 'we humans'. 'We the people' are always defined by a place within humanity, both social and geographical. So there is a distributional question: who gets what environment – and why? As to 'we humans', there are qualities we share as a species, and we humans have now to consider our relationship with the non-human world. The struggle for justice as it is shaped by the politics of the environment, then, has two relational aspects: the justice of the distribution of environments among peoples, and the justice of the relationship between humans and the rest of the natural world. We term these aspects of justice: *environmental* justice and *ecological* justice. They are really two aspects of the same relationship.

SITUATING JUSTICE

We take an approach which situates the discussion of justice in actual events. Many texts on environmental ethics begin by posing questions which assume a societal, or even global, frame (Hardin, 1968; Commoner, 1972; Drengson, 1980; Tokar, 1987; Spretnak and Capra, 1986; Sessions, 1989). These 'big questions' commonly address: the capacity of the earth's resources to support its human population; the capacity of the biosphere to absorb human wastes; climate change as a result of human agency; the rapidly increasing rate of extinction of non-human species; the exploitation of the environment of the poorer nations to maintain the lifestyle of the richer; the systematic discounting of the interests of unborn generations; the massive injury to the forests and seas; and the industrial use of animals. Conclusions are then drawn about the kind of society and morality we have to develop to prevent these things happening: the society of 'our common future' (Brundtland Report, 1987).

Ultimately, political and environmental ethics must address this 'big picture' because so many ecological and social problems have a systemic or structural basis. We need political–ethical frameworks which can help humanity to address those threats which it faces collectively. Nevertheless, if the struggle for justice is a real world process then we must make clear how abstract conceptions are connected with real world events. Threats to the environment, however global, are manifested in specific places and local contexts. If society is to change to accommodate new conceptions of justice, it is necessary to demonstrate in an immediate and concrete way why the existing means of dealing with environmental conflicts are inadequate. Social change on the scale which may well be necessary for global society to carry on the task of finding and delivering justice in and to the environment is likely to proceed in a somewhat piecemeal and incremental way. However, incremental change, as we have seen repeatedly in this century, can have far-reaching consequences (Swyngedouw, 1992).

Our point of departure, then, is not the big picture of 'our common future', but examples of actual and public conflicts over the environment. The broader struggle against social forms which produce environmental injustice, biospheric destruction and the maldistribution of environmental risk begin as engagements with specific conflicts.

In 1995 three incidents occurred which tell us something about the nature of environmental conflict. These incidents are a small and not necessarily representative sample of the many reported and unreported cases of exploitation which, taken together, present a threat to the ecological integrity of the planet. They are the kinds of issues on which, in one way or another, judgement is passed. The first was the unsuccessful attempt by the Anglo-Dutch transnational corporation, Shell, to sink one of its obsolete oil rigs in the North Atlantic; the second was the conduct by France of a series of underground nuclear tests in Pacific atolls; the third was the mining of metallic ores by the Australian transnational corporation Broken Hill Pty Ltd (BHP) in Papua New Guinea.

Disposal of the 'Brent Spar'

When oil-drilling rigs become obsolete their owners must find a way of disposing of them. The first Norwegian rig to be decommissioned had been brought ashore for dismantling. But Shell, the joint owner with Exxon (Esso), decided it would dispose of the British Brent Spar rig by towing it out into the North Atlantic and sinking it in deep water. Shell obtained approval from the British government for the dumping. But there are many oil rigs around the world which are reaching the end of their life, and the principle of dumping at sea was vigorously opposed by the green movement in Europe. The rig is a steel and concrete tube the size of an upended aircraft carrier. The main environmental hazard is allegedly caused by the oil waste and radioactive scale contained in the tanks. It was feared these poisons would eventually seep out into the sea.

The international activist organisation, Greenpeace, launched a political campaign including a consumer boycott in Europe aimed specifically at Shell. The main focus of the campaign was Germany, Denmark and the Netherlands. Governmental action in Europe was mobilised around the Oslo and Paris Conventions which regulate the disposal of waste into the North

Sea. The 1972 Oslo Convention covers the prevention of sea pollution by dumping from ships and aircraft, and the 1974 Paris Convention covers the prevention of marine pollution from land-based sources. These Conventions were drawn up within the framework of the International Maritime Organization (IMO).

The consumer boycott had an immediate effect on Shell's fuel sales, which slumped by 30 per cent, and the company was dismayed by the tarnishing of its carefully cultivated image of environmental responsibility. This image was, however, finally ruined in late 1995 by the publicity given to the company's involvement with the Nigerian government in the violent suppression of protest against Shell's environmental degradation of the land of the Ogoni people of Nigeria.

Intergovernmental action in Europe led by the German government, rather than the consumer boycott, was probably in the end decisive. The German government, applying the precautionary principle, insisted that the risk was substantial (Johnson and Corcelle, 1989: 294–5 and 301).

On 21 June, at the very moment that the British Conservative Government was stoutly defending Shell in Parliament, Shell capitulated to the political and consumer pressure and the tugs towing the rig turned around, providing a powerful visual symbol of victory for the green campaign. Greenpeace won this battle, but the vast oil rig still has to be disposed of. It is true that disposal on land may render the process more open to scrutiny, but whether the process will create any less pollution of the air, soil or water remains to be seen. Even if the process of disposal is potentially open to public scrutiny, it will most probably not attract the media attention brought to bear on the single dramatic event of a sinking at sea. The focus of the issue was the *distribution* of risk rather the *production* of risk (Lake and Disch, 1992; Dryzek, 1987), a critical ecological distinction which we will address in Chapter 5.

Paradoxically the great political triumph of Greenpeace was later acknowledged by that organisation to have been of doubtful value to the environment. In the rapid mobilisation of opposition, accurate information was in short supply. The *representation* of risk became the principal objective of those opposed to the dumping. Greenpeace retreated from its earlier position that marine disposal of the rig posed a threat to the environment. Whatever the actual risks involved may or may not have been, they were never subjected to careful examination and judgement. Winning the battle became an end in itself.

French nuclear tests in the Pacific

On 7 September 1995 the French government carried out the first of six nuclear tests on Mururoa and Fangataufa atolls in the Pacific. The French Prime Minister said, 'the nuclear tests will incidentally (*sic*) not have any impact on the environment because they are carried out at very great depths in solid rock'. Other experts disagree. The French newspaper, *Le Monde*, has published photographs reportedly taken by French divers in Mururoa lagoon in 1987 which show cracks three metres wide and several kilometres long in the volcanic structure. Pierre Vincent, the French vulcanologist, considers it possible that a flank of the basalt rock below water level could shear off and fall into the sea. The evidence suggests that such shearing is a normal event in volcanic explosions as was demonstrated when Mount St Helens exploded and the north flank of the mountain broke away. Computer simulations carried out in New Zealand suggest that radiation may already be leaking into the sea through cracks in the basalt, and will in any case do so some-time in the next hundred years (Cookes, 1995). Dr Tilman Ruff, a physician at the International Health Unit of the Macfarlane Burnet Centre for Medical Research, Melbourne, writes that there is:

> clear evidence, documented by the brief and limited independent scientific and medical missions that have visited French Polynesia, of extensive damage to Mururoa; venting and early indications of long term radioactive leakage from underground nuclear explosions; well documented outbreaks of ciguetera fish poisoning caused by the nuclear test program; and adverse social effects of nuclear colonialism.
>
> (Ruff, 1995; see Figure 1.1)

Whether or not actual damage to the atolls has yet occurred, whether or not the islands are yet leaking radiation, there is a very definite risk, indeed a probability, that radiation will escape in future. Ulrich Beck (1995) has observed that the discourse over the environment is primarily a discourse of risk. The experience of a nuclear accident at Chernobyl has alerted the world to the devastating and widespread effects of a nuclear leak into the environment. Even if the risk is small, the disaster risked is enormous. It is all the more shocking that the military nature of the test programme makes adequate public scrutiny of the risk impossible. No sanctions are available which would make the national interest coincide with the global interest. The evidence cannot emerge to provide any test of truth about present damage. Protests from other nations, the opinion of French people, and pressure from non-government organisations had little immediate impact. The French government stopped testing when it suited the French government.

Of course, reasons of state supplement reasons of the economy. France's continued development of nuclear weapons supports a nuclear industry situ-ated in Monsieur Chirac's particular constituency, the Paris region (Chirac was the first Mayor of Paris, and still occupies that position in addition to the presidency). The nuclear tests were to result in the final certification of the warhead of the M 45 multi-warheaded missile carried by a new genera-tion of nuclear submarines, the SNLE-NG (Ruff, 1995). Australia, while

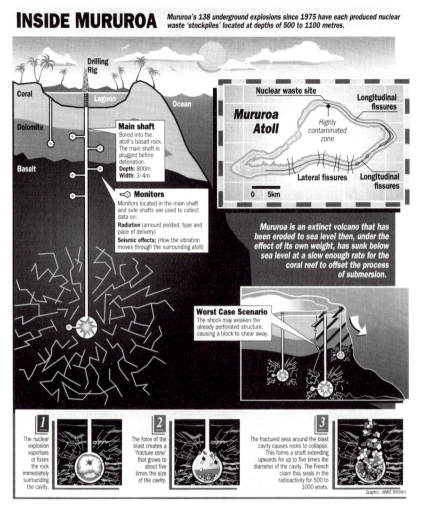

Figure 1.1 French nuclear tests: inside Mururoa Atoll

Source: Detail from a diagram by Jamie Brown, published in *The Age*, Melbourne, 12 August 1995

protesting loudly in public, sells uranium ore to France which is used in the French nuclear reactors which produce the fuel for the nuclear devices.

Mining in Papua New Guinea

Since the mid-1970s the Australian mining corporation, BHP, in close co-operation with the government of Papua New Guinea has developed one of the biggest open-cut copper and gold mines in the world. The Papua New Guinea government has taken a 30 per cent equity stake in the mine. The

6

mine is at Mount Fubilan in the northern mountains near the source of the Tedi river (Ok Tedi). Ok Tedi is part of Papua New Guinea's biggest river system, the Fly River, which flows into the Gulf of Papua. The mine has gouged an enormous crater in the mountain, destroyed the local rain forest and discharges about 80,000 tonnes of limestone sludge per day into the upper reaches of the Ok Tedi (see Figure 1.2). The sludge contains many chemicals and minerals including copper particles in concentrations of up to 18 per cent of the waste.

Since mining began in the mid-1980s, over 250 million tonnes of waste have been dumped into the river. During the wet season, when river levels rise very quickly, an impervious blanket of mine sediment is deposited on the forest floor downstream. In places this blanket is more than a metre thick. For about thirty square kilometres along the river flood plain, the forest has died. The Government of Papua New Guinea has admitted that the environmental damage cannot be repaired (according to Mines Minister Iangalio as reported in the *South East Asia Mining Newsletter* 10 Sept 1993, p. 5). The mining will continue for at least another fifteen years and over a billion tonnes of sludge will be dumped.

The environmental damage is large, almost certainly irreversible and largely unpredictable. The ecology of the Fly River system has been changed by the sediment, and if claims that the river system is *now* biologically dead are exaggerated, the future effect of the dumping is largely unknown. There has been no independent environmental monitoring of the environmental damage. The only assessment is done by consultants paid by the mining company. It is known, however, that fish have vanished from some parts of the Ok Tedi, that the area of rain forest dying from the sediment will extend much further along the river flood plains as mining continues and that the bed of the river has been raised by more than a metre of contaminated silt over seven years.

In 1984 the mining company (Ok Tedi Mining Ltd) tried to build a tailings dam to contain the sludge but the dam collapsed in the course of construction and the company gave up the effort. BHP claims that a tailings dam is impossible to build in the geologically unstable terrain with an annual rainfall of up to 10 metres. Moreover an unsafe dam would of course pose the threat of catastrophic flood to the 30,000 downstream villagers. But there is no doubt that the villagers have already suffered. Their river-bank gardens have been destroyed, fishing has become impossible and the wild boar they used to hunt have disappeared.

The project has enormous significance for the Papua New Guinea economy. The nation was the sixteenth most indebted in the world in 1991. The debt stood at 130 per cent of GDP (*The Economist*, 1993). A high level of exports is therefore essential to pay the annual interest bill. The Ok Tedi copper mine accounts for at least 16 per cent of the country's export earnings. The

Figure 1.2 The Ok Tedi mine, Western Province, Papua New Guinea

Source: Photograph by Bruce Miller, January 1990, published in *The Age*, Melbourne, 8 February 1990

after tax profits from the mine for the year 1994–5 were $A250.9 million (about £115 million). The company has provided compensation to some of the villagers, to the value of $A13 million, in the form of 'meeting halls fresh water and shower blocks' (Skelton, 1995). The villagers, however, have lost the entire environment which supported their way of life. Villagers in the area far downstream of the mine, who were not included in the original mining agreement, claimed $A4 billion in compensation (see Figure 1.3). A compensation package was offered to the villagers of $A110 million over the life of the mine.

A Melbourne law firm, Slater and Gordon, took the case for compensation on the basis of a 'success fee' and sued the company in the Australian (State of Victoria) and Papua New Guinea courts. But the company has been able to combine its economic power with the coercive power of the Papua New Guinea state. A law has been drafted with the help of BHP to make it a criminal offence for the villagers to seek redress through the courts. This is, of course, outrageously unconstitutional and tears up the rule of law. The law firm was harassed and denied access to their clients in Papua New Guinea. In Australia there remains some respect for the principle of 'separation of powers', at least in judicial circles. The lawyers acting for the villagers brought an action in the Supreme Court of Victoria against BHP for contempt of court, and a contempt finding was duly delivered on 20 September 1995. The court ruled that the suit against BHP could proceed and that a law introduced by the Government of Victoria to prevent such contempt actions was unconstitutional. The Government of Victoria, however, altered judicial arrangements to transfer the power to initiate contempt proceedings from an independent public prosecutor to itself. The Government through the Attorney General of Victoria declined to proceed against BHP for the contempt. In an extraordinary later development it was revealed that at the time of the decision the Attorney General held 710

Figure 1.3 Pastor Maun Tepki with his people at Bige on the Ok Tedi river
Source: The Age, Melbourne, 23 September 1991

9

ordinary shares in BHP worth over $A12,000 (*The Age*, Melbourne, 18 October 1996, p. 1)

Since German firms were among the partners in the mining venture, the case attracted considerable publicity and criticism in Germany. Eventually, in 1994, the German partners withdrew from the venture. BHP also suffered severe damage to its carefully cultivated public image as a good and environmentally conscious corporate citizen. Richard Jackson, Professor of Geography at James Cook University in Northern Queensland describes the 'mine of disinformation' surrounding the Ok Tedi dispute in which both sides used the media to present their case (Jackson, 1995). Jackson writes, 'No-one in Australia can fail to be aware of the basic issues: uncaring, Australia-based companies have sought huge profits in Papua New Guinea while impoverishing traditional village communities and destroying a previously pristine environment'. An out-of-court settlement was reached in 1996.

Ok Tedi is not an isolated case in Papua New Guinea. The Panguna mining operation on Bougainville Island has been held up for ten years by armed opposition by local people, and in February 1997 the Prime Minister, Sir Julius Chan, announced that mercenary troops would be deployed to end the opposition. This act plunged Papua New Guinea into civil violence, armed insurrection by the military and a constitutional crisis. In the subsequent general election, Sir Julius lost his seat. Ok Tedi presages a flood of mining projects which are just waiting for a new surge in global economic growth. Baker reports:

> Capitalising on their proximity to Asia, our globally focused resource groups, adventurous junior explorers and deal-seeking entrepreneurs are all leading the rush by western countries into the highly prospective region. Australian companies are in the vanguard of the vault into Indonesia, the Phillippines and Vietnam, and are pioneering pushes into Myanmar (Burma) and Mongolia. They are striding into China and India and getting their feet wet in such places as Malaysia, Thailand and Pakistan.
>
> (Baker, 1996: 6)

He continues: 'Speeding the move into Asia is a range of perceived impediments to doing business at home, such as native title legislation, environmental regulations, high taxes and the fact that there are few places to go exploring' (the native title legislation recognises the first occupancy of the continent by indigenous peoples and the seizure of their land by European settlers without treaty or contract). So there is a double standard for environmental and ecological justice for Australia *vis á vis* the countries of Asia. The next project, which will begin within two years, will be the billion dollar gold mine at Lihir island off the Papua New Guinea coast. This project is the child of the vast mining corporation formed from the merger of CRA and Rio Tinto Zinc Ltd. (RTZ). The Lihir project will excavate the crater of a collapsed volcano on a small island off the Papua New Guinea coast and discharge the tailings directly into the sea. If the gold mine uses the usual cyanide process, the tailings will be extremely toxic and the marine environment for many square kilometers around the site will probably be devastated.

Environmental degradation in East and West

There are many other examples of what might be called environmental insults (in the medical sense of damage to the body) in West and East. In Guyana, at the second largest open-cast gold mine in South America, the wall of a holding pond collapsed. This pond contained cyanide-laden effluent. The mine is run by Omai Gold Mines Ltd and 65 per cent is owned by a Canadian multinational Cambior Inc. The diluted cyanide (at 15 parts per million, seven times more deadly than the accepted 'fatal dose') flowed into the Essequibo river system and immediately killed everything in its path. The government of Guyana declared a state of emergency as the cyanide threatened human life around the river.

Dramatic single incidents such as the Chenobyl disaster, devastating though it was (see Medvedev, 1992), by no means represents the worst of the results of socialist productivism. As early as 1980 a samizdat tract denouncing pollution in the Soviet Union was published in the West (Komarov, 1980). Wolf (1992) writes of the massive ecological degradation of the land. Islands of pollution are now merging to form a block of ecological degradation streching from the Baltic to the Black Sea. Reznichenko (1989, cited in Kapuscinski, 1994) attributes the catastrophe of the drying up of the Aral Sea to the introduction of an irrigated cotton monoculture along the two great rivers, the Syr Darya and Amu Darya, which once fed it. Nor can it be imagined that the Soviet Union's successor has 'turned green'. Certainly Russians are now free to form their own environmental movements and pressure groups, but a powerful and intransigent bureaucracy remains, whilst the growth of market relations threatens Russia's environment in new ways. Both the environmental damage which occurred under state socialism and the new dangers in the transition to capitalism in the former Eastern bloc countries of Europe are described in Carter and Turnock (1993).

There are many less dramatic events occurring every day which pose just as serious but less direct threats to the environment. In Australia in 1995 the State of Victoria sold off its electricity distribution system to private operators from around the world. The sale of the state distributor Eastern Energy (one of five distribution corporations) to Texas Utilities for $A2.08 billion (about £940 million) was Australia's biggest public asset sale. The price was considered very high, and the following day Australians learned why. Mr Dan Farell, the new chairman of Eastern Energy, said he hoped to boost profits by increasing the sale of electricity. Farell said Victoria's electricity consumption was low compared with Texas. Eastern Energy customers use, on average, 5,600 kilowatt-hours per year compared with 14,283 kilowatt-hours in Texas. Every additional kilowatt-hour, however, adds to Australia's greenhouse gas emissions from the brown coal burned to generate the electricity (see Walker, 1995a, b). In the same year the government of Victoria also decided to build an extensive addition to Melbourne's freeway system. The freeways are to be financed by private investors who will draw revenue from tolls on the roads. The road system can only deliver a financial return to investors if there is a 30 per cent growth of road traffic and a further decline in the use of public transport. Both these projects tie private profits to a form of growth which will increase greenhouse emissions. The Australian Government now wants the world to treat Australia as a special case and waive the greenhouse targets that the rest of the developed world has voluntarily adopted.

The most basic choice we confront is between maintaining the quality of the planetary environment, and its exploitation for the production of commodities for human use. For a time it seemed that the moral choice could be avoided by applying the formula of 'sustainable development'. Joining two positive-sounding words seemed to resolve at a stroke the conflict between an economy based on everlasting growth and a planetary environment of permanent high quality. These goods, it was hoped, could be reconciled if only the economy could be organised around production activities which did no harm to the environment. According to the Brundtland Report (1987), sustainable development is: 'development that meets the needs of the present without compromising the ability of future generations to meet their own needs'. Can conflicts of the kind described be satisfactorily resolved under such a rubric?

SUSTAINABLE DEVELOPMENT, ENVIRONMENTAL RISK AND ECOLOGICAL MODERNISATION

The hope of 'sustainable development' is based on the observation that not all 'development' is environmentally degrading. The early environmentalist goal of 'zero growth', if it means no more than zero growth of environmentally depleting activity, will not be sufficient to prevent long-term damage to the biosphere. Jacobs (1991: 57) discusses the example of the tropical rainforests. Even if there is zero growth in the rate of destruction of the forests, the habitat provided by these forests will still be destroyed, most probably irreversibly, within the forseeable future. Not zero growth but zero destruction is required here. On the other hand, even *increased* economic growth in some industrial sectors may be entirely compatible with environmental *improvement*. The growth of new industrial sectors serving the needs of cleaning up existing industrial processes is one example.

One can argue over the definition of 'development' and 'sustainable' but what seems undeniable is the fact that judgements must be made in favour of some forms of development and against others, and further, that those judgements can no longer be left to individualised producers and consumers interacting in markets created and conditioned by national states.

To take one example close to the authors' home, there are two timber industries in Australia. One industry logging native forests has been established for over a hundred years. Families of loggers who have grown up in the rural environment are supported by this industry (see Figure 1.4). A more recent profitable addition is the conversion of surplus trees into woodchips for the Japanese paper industry. These 'surplus' trees, however, are the habitat of many animals. The other timber industry is the intensive cultivation of plantation timber. Pine plantations have been cultivated for many years for production of softwood for building timber. More recently, however, agricultural land has been planted with fast growing gum trees (eucalyptus) to supply the home and export woodchip market (Figure 1.5). Some Australian paper mills have now planted eucalyptus to keep them well supplied with the raw material for pulp into the future. Plantations of eucalypts are arguably better for the land than the European-style agriculture that the plantations replace.

One industry is damaging to the environment, the other is not, and will probably improve it. The plantation timber industry has the capacity to supply all existing demand both for logs and woodchips within a few years. Logging and woodchipping of ancient forests is in one sense 'sustainable' because the trees are replanted, as required by law. But virgin forest is slowly being converted into plantation forest. In a very important sense this practice is unsustainable because it destroys the habitat of rare species which live in and around the old and often decaying trees. Traditional forestry practice, which involves massive human intervention, is not necessarily good conservation practice, and the competition from the conversion of virgin forest to plantation forest inhibits the growth of truly sustainable plantations on agricultural land. For the past several years the Australian government has devised strategy after strategy to allow logging and woodchipping to continue, while trying (unsuccessfully) to placate the environmentalists with patchwork plans for forest conservation, which show an overall 'bottom-line' increase in land protected. Environmentalists remain unimpressed because the government continues to include areas already logged within the 'protected forests' (see Figure 1.6).

The issue illustrates the complexity of 'sustainable development'. The logging industry is subsidised in various ways by the governments which own the forests and act to support a dying industry. If the plantation industry were allowed to compete on an equal footing with the logging industry, the outcome might be better for the environment. Yet, a pure market solution, even if such a solution could in practice be defined, is by no

Figure 1.4 Woodchipping near Daylesford, Victoria, Australia
Source: Photograph by P. C. Sillitoe, 23 December 1994, published in *The Age*,
Melbourne, 4 February 1995

means guaranteed to produce such an outcome. Sustainable development, as Jacobs (1991) has argued, involves conflict of interest between different industries and those who depend on them; it involves conflict of interest between developed and developing nations, and it involves conflicts between the interests of generations. These conflicts demand just solutions. Sustainable development without environmental justice is an empty formula designed, in Jacobs's words, to 'wave away such conflicts in a single unifying goal' (*ibid*.: 59).

Science cannot by itself provide solutions to such human conflicts. But it does have an important forensic role. In modern society, Beck (1992) argues, the production and distribution of wealth is accompanied by the production and distribution of 'risk'. Risk, as Beck defines it, is not something new. The agricultural exploitation of nature has always been accompanied by the risk of overexploitation, eventually inducing famine. But, as the transformative potential of modern industrial production has unfolded, we have become increasingly aware of its widespread effects. This awareness is also accompanied

Figure 1.5 Plantation timber – Bob Baxter of Karadoc winery
walks through fast-growing gum trees
Source: Published in *The Age*, Melbourne, 17 April, 1995

by the knowledge that we cannot know precisely what these effects will be. The more we know, it seems, the more we become aware of the limitations of knowledge. Processes and substances thought harmless are revealed, within the space of twenty years, to be life threatening (note some of the examples discussed by Wenz, 1988).

In the distribution of risk there is a parallel with the distribution of

Figure 1.6 The Australian Prime Minister reissues licences to
export woodchips from old growth forests

Source: Cartoon by Ron Tandberg, *The Age*, Melbourne, 28 January 1995

wealth. Modern production spreads both and also concentrates both, as has been noted above. The possibility of protecting elites from the risks of modernisation is shown to be limited – this is not new. But the special characteristic of 'risk' is its dependence upon *uncertain* knowledge. That science offers *only* uncertain knowledge, that is propositions, hypotheses, probabilities, has been understood by scientists for a long time, but this understanding is now a public matter. Debates have always been part of science. But scientific debates have increasingly entered the public political arena. It is understood that the debates take place and that their outcome affects everyone. So the production of scientific knowledge has escaped from the laboratory.

What are the implications of this escape? Beck points to three major effects. First, risks only exist 'within knowledge'. They are based on causal interpretations and are open to social definition and construction: 'hence the mass media and the scientific and legal professions in charge of defining risks become key social and political positions' (Beck, 1992: 21). The struggle for the environment is fundamentally a political struggle for control over public knowledge in which the traditional legitimacy accorded to 'science' is fast disappearing. Second, the boundary between scientific discourse about things, and the discourse about the meanings of these things to human beings is dissolving. All scientific knowledge of risk becomes essentially forensic science functioning as evidence entering a political process.

There is conflict not only between bodies of 'evidence' but between the principles on how to treat them. In actions which affect the environment the German 'precautionary principle' (*Vorsorgeprinzip*) emphasises the burden of proof of environmental *safety* (of the action), while the English principle emphasises the scientific burden of proof of *unsafety* (see Weale, 1992: 79–81). We can see something of that difference underlying the attitudes of the German and British governments to the proposed sinking of the Brent Spar oil rig. Third, risk, through the dissemination of knowledge, becomes the target of the production system. If the alleviation of risk is something that people demand, then this demand can itself be manipulated by the creation, through public knowledge, of 'risk' awareness.

It is obvious, then, that the production of knowledge about the environment plays a pivotal role, not only in the production of material goods and material effects like pollution, but in the production of modern society itself. In the growing experience of risk, in Beck's words, there is a normative horizon of lost security and broken trust:

> Risks remain fundamentally localised, mathematical condensations of wounded images of a life worth living. . . . Behind all the objectifications, sooner or later the question of acceptance arises and with it the old question: how do we wish to live? What is the human quality of humankind, the natural quality of nature which is to be preserved?
>
> (Beck, 1992: 28)

Of course such a question begs further questions such as who is this 'we', whether the lives of 'we's conflict, whether the humanness of humankind conflicts with the naturalness of nature, and how conflicts among different 'we's and 'them's are to be settled These questions cannot be answered without reference to ethical standards which necessarily contain the matter of justice.

In the coming century capitalist development will bring the newly industrialising world up to the standard of consumption of the old post-industrial world. That will mean an increase in global production which will dwarf that of the industrial revolution of the nineteenth century. There is today little public conception of the impact of that growth on nature or on humankind. Looking back from the perspective of the year 2100, we will perhaps see that the development of the nineteenth and twentieth centuries was in fact only a miniature precursor of the vast wave of industrialisation which engulfed the planet in the twenty-first century. It would be a mistake to suppose that the violent means of resolving conflicts so characteristic of the twentieth century were some kind of aberration. It will be astonishing if the world reaches the twenty-second century without nuclear war, without an increase in death camps, genocide and mass starvation and with some semblance of democratic governance.

In these circumstances the institutional measures to modernise economies in line with ecological imperatives and to control pollution seem pitifully inadequate. As Weale remarks:

Environmental problems . . . are now present on a scale and in a form that dwarfs previous experience of legislation and policy making. There is no reason to believe that the political institutions of modern democracy are capable of responding quickly and effectively to these problems, any more than they have been capable of responding to the mass poverty and unemployment of the inter-war period, themselves novel problems that governments failed properly to understand or to solve in the 1930s.

(Weale, 1992: 145)

In those times as in ours the more far-sighted politicians and academics began to re-examine the moral content of policy. The blatant injustice of the social conditions created by existing policies had to be addressed then and must again be confronted. The present prosperity and future economic prospects of a poor nation are portrayed as depending on environmentally degrading industry being allowed. The question tends to be framed in terms of an opposition between the national economy and the environment (national, global, local). Lipietz (1992), however, formulates the question differently. He asks: 'Is the conversion of the vast surpluses of global corporations into debt, which chains countries to competitive environmental exploitation for the purposes of creating more capital for the global corporations, really in the best interests of world development?'

The need for justice is by no means a one-sided matter. Of course, there are contradictory interests and pressures but institutional frameworks are devised to resolve these contradictions in a just manner. A just institutional framework for the conduct of business is one which is transparent, predictable and equitable, one in which the same rules apply in the same way to all corporations competing for business within an economic sector, in short the rule of law. Arguably it is the absence of such conditions, a chaotic situation maximising uncertainty, which exacerbates conflict and makes it harder for corporations to do their job effectively. Corporations will compete as best they can to pursue their short- and medium-term interests in the political conditions they find. They cannot be expected to set about changing those conditions. This is a task which has to be addressed collectively. The problem today is that the institutional means for tackling it is inadequate.

The question of global justice was posed in the Rio Declaration of 1992. Lipietz comments:

In the same way that social democracy went beyond civil democracy, political ecology appears like an *aufhebung* of social democracy: the recognition, at first moral, of new rights, of new bearers-of-rights, and of new objects of rights, thus of obligations and of new interdictions. Perhaps the greatest advance of the Rio Conference and of its hundreds of parallel conferences, will be to have solemnly – and in the

18

mass media – recognized new rights and obligations to be incorporated within social norms in accordance with honesty, respect of the other, or the Universal Declaration of Human Rights. Rio has built the basis of jurisprudential justice, even without positive and democratically established international legislation.

(Lipietz, 1996: 223)

The impacts of environmental degradation are always socially and spatially differentiated. They may end up affecting the global environment, but first they damage small parts of it. These local effects frequently transgress national boundaries – acid rain from England damages Danish and Norwegian forests, water pollution from Switzerland destroys the ecology of the Rhine in Germany and the Netherlands. Then international distributional questions reach the public arena and international law becomes involved (see, for example, Wenner, 1993).

However important is the question of the distribution of environmental goods between present and future generations, a seemingly simpler choice has yet to be confronted: the just distribution of the wealth of nature and environment *within* the population now living. Do we think that the present distribution of wealth and power to consume is morally correct? Or do we prefer to think that such distribution is not a moral question? Because the market is a useful mechanism for serving individual tastes, do we suppose that all distributional questions are reducible to a matter of tastes? It seems more than a little inconsistent to show moral concern for future generations when the worst environmental conditions imaginable are already present in places on this planet today. A United Nations report in 1993 said baldly that the world environment had deteriorated irreversibly over the past twenty years (United Nations, 1993). Within a shrinking biosphere the question of what is sustainable will increasingly be a matter for political debate as well as scientific assessment. Judgements are becoming unavoidable, and the need for action urgent.

Distributional questions are fundamental to the politics of the environment. The question of justice *within* the environment is enfolded in the question of justice *to* the environment. All the actors involved have interests in pieces of the environment. Proximity is at the heart of the struggle. The quality and perhaps the meaning of each person's existence depends upon a 'world' that wraps around the person, an *Umwelt*, which is neither bounded by the institutional barriers of property and the territory of states, nor is infinite in extent. The division of land into property and territory may be convenient from the point of view of economic exchange, but it is arbitrary from the point of view of environments. Environments overlap and are *unavoidably* shared. The sharing extends both to the 'goods' and the 'bads' they contain. Is this division ultimately founded on anything more than the 'right' of the more powerful? If so how can this 'right' be modified to give due recognition in our ethics and politics to the fact of the shared and collective experience of the environment?

This book is not primarily about the means but about the ends of institutional change. The means, however, cannot simply be ignored in such a discussion. The question of what is possible must always form the background for debates about what is desirable. New institutional structures and processes can be developed, and have already been invented and tested in different national contexts as is shown in Weale's (1992) discussion of the Netherlands Environmental Policy Plan, the Swedish integrated pollution control system, the German precautionary principle and innovative institutions for resolution of environmental conflict tried out in the US. More radical options were canvassed by Dryzek (1987). We have to hope that a sense of urgency can be found to expand this invention and testing. Those who argue for a more revolutionary approach should heed the warning of March and Olsen (1989: 65) about the outcomes of intentional transformation through 'radical shock': 'It is easier to produce change through shock than it is to control what new combination of institutions and practices will evolve from the shock'.

Of course capitalism is unjust. Any world infused with power is unjust, human nature being what it is. Unfortunately a world without power is a very long way off. Nevertheless, if the best we can hope for is the correction of *injustice*, this still requires that we conceive of *justice*. We do not need to – and indeed must not – give up our grand narratives, but let us acknowledge that we can only make progress step by step, feeling our way as we go. Thousand year reichs, giant leaps forward, cultural revolutions, twenty-year plans have not been marked by much progress for humanity or nature in this century. A little caution must be applied in the next. A new society, a new economic system cannot be instituted by a revolution or a stroke of the pen. It will have to evolve with the consent of the world's peoples. The question is, what institutional changes are likely to lead in a progressive *direction* of systemic change – and give rise in their turn to *further* systemic changes in that progressive direction?

The twentieth century is a century of global scale: global war, global economy, global environmental hazard. But our politics remain place-bound, and specifically nation-bound. Our most effective global institution, the multinational corporation is situated for the most part outside politics, mostly beyond the scope of law, and with only a non-compulsory attachment to ethical behaviour. The nation state, once the provider of a framework of law for the corporations, may turn out to be little more than their junior partner in environmental exploitation. The concern with 'sustainable development', or worse, 'sustainable management' is indicative of our neglect of justice. The dialectic of justice has reached an impasse in which the struggle over ideas – though present in abundance – has come to have very little effect on real human–human and human–nature relations.

So we must return to justice. But the philosophy of justice is fraught with disagreements, tensions, antinomies and contradictions. We do not shy away from these disagreements. Nor do we regard them as a sign that the philosophy of justice has 'failed' because it does not deliver a single perspective on which everyone agrees. On the contrary, we view such disagreements as a sign that the philosophy is a live human process, a process which is actually definitive of humanity, a *dialectical* process in which many people struggle to define the truth about justice, a process of 'finding justice'. It is this process we need universal principles to protect.

THE DIALECTIC OF JUSTICE, SOCIETY AND NATURE

Dialectic, Bhaskar explains:

> In its most general sense . . . has come to signify any more or less intricate process of conceptual or social (and sometimes even natural) conflict, interconnection and change, in which the generation, interpenetration and clash of oppositions, leading to their transcendance in a fuller or more adequate mode of thought or form of life plays a key role.
>
> (Bhaskar, 1993: 3)

Dialectic, thus concerns the play of opposing *ideas*, opposing social *agencies*, and opposing natural *forces*. Let us now sketch out the dialectic of justice, society and nature as we shall try to unfold it in this book.

We seek to answer two main sets of questions about our human relationship with the rest of the natural world. First, do humans have a claim to fair shares of the particular good which 'an environment' provides? What is *environmental* justice? Second, does non-human nature, or at least aspects of nature, have a claim on justice? What is *ecological* justice? We will argue that these two sets of questions must be addressed together within one problematic, whose solution lies in a political ethic of justice.

The cases discussed above embody the contradiction between 'development' viewed as growth of business activity (GDP) and environment. In each case 'the environment' is both a local *Umwelt* and part of '*globus terraqueus*', our fragile planet of which we ourselves are an integral part. Each is a case for judgement. In these examples the dialectic of justice takes the form of a material struggle in which the criteria of justice conflict: the rights of company workers, the rights of those affected by a company's work in different ways, the rights of indigenous people, the rights of non-human nature (to instance just a few of the conflicting principles which could be brought to bear on the situation). In each case there is as yet no 'rule of law' which could give security to the parties involved in anticipation of the judgement which would be made in the light of all the facts and ethical principles relevant to the case.

21

Action to solve the world's environmental crisis must start by being able to deal justly with the local acts which are destroying the environment on a daily basis and as a matter of course. The pursuit of environmental objectives exclusively through pressure-group politics has reached an impasse. We argue for a new rule of law within which the world's corporations can conduct business with the security that, in doing so, they are not destroying the planet.

The basis of such a rule of law we begin to investigate in Chapter 2. Both ethics and justice have been backgrounded – tending to disappear – in some important philosophical traditions relevant to society and nature. After being banished by certain neo-Marxist perspectives, we show that an environmental ethic was embodied in Marx's thought. We argue, moreover, that Weberian rationalism, however 'ecological', does not dispose of the need for ethics. Ethics has to be part of the dialectical struggle, part of the wrangling with ideas in order to understand our social, natural and ecological situation.

If indeed ethics is necessary to our social organisation and governance, is a *justice* ethic necessary? Justice is in danger of disappearing in the substitutions demanded by communitarians and some post-modern feminists. But the context of global communication and power provided by the world economy structures the relations among communities and limits their capacity to develop autonomous ethical systems. It is the justice of the global context that must now come under critical scrutiny. The relativist tendency in feminist and communitarian thought and the concurrent failure to discriminate between occasions when difference must be fostered, when tolerated and when opposed leads to abdication from critique. To engage in critique, we argue, necessitates the reinstatement of universal principles of justice – albeit at a second political level.

The ethic of 'care' is too weak a principle for dealing justly with the 'other', either in humanity or in nature. Care can too easily be viewed as dependent on individual empathy and thus become optional. *Caritas*, without justice, can too easily become charity. Justice must be necessary and universal. But the ethic of 'inclusiveness' is an essential principle of the dialectic which brings both the excluded of humanity and the non-human world into our moral consciousness. Following Arendt and Benhabib we argue for 'enlarged thought' in a vigorous democratic politics. The introduction of new ecological perspectives into the dialectic increases the potential for 'enlarged thought' but at the same time demands new institutions to enable this enlargement to have purchase on political reality.

If we accept that justice must be at the core of our philosophy, then what are the bases of justice? These have been correctly identified by Miller (1976) as 'desert', 'rights' and 'needs'. In Miller's view these bases constitute irreconcilable ethical systems. We examine the bases of justice in Chapter 3 and find that there is a good deal of overlap between them. The words 'deserve', 'has a right to', 'has need of' express different aspects of what is entailed by a single underlying principle: that of human autonomy. Likewise the dialectical relationship between local culture and the totality of the human species (sharing bodily fragility and specifically human consciousness) runs through the interpretation of the bases of justice.

The ways in which the bases of justice have been combined in different ways create first-level systems of *political* justice. These we examine in Chapter 4. The ethical foundation of global capitalism today is a narrow and selective combination of aspects of utilitarianism and property entitlement. The consequentialism of the utilitarian perspective is indispensible to the development of environmental and ecological justice. Yet, the way in which measurement of a certain restricted kind, namely the measure of business activity, has become inscribed as a generalised measure of 'success' for everyone and everything everywhere grossly distorts the consequentialist principle, reinforces unbalanced power relations in the world economy and creates an artificial dichotomy between development and conservation. The limits of science and measurement are also the limits of a consequentialist ethic if pursued to the exclusion of deontology.

In the defence of 'property' from Kant to Hayek to Nozick, 'entitlement' has become more and more circumscribed until it is no more than the right of the powerful in a catallactic, chrematistic plutarchy, which is the world economic system. Yet, the ideas of 'property' and 'entitlement' contain the elements of an important ethical principle which, as with desert, rights, and need, finds its basis in the principle of human autonomy.

In Kant's view all people are equally entitled to property. The fact of inequality is thus what has to be explained and justified in contractarian theory. For Kant the contract was an ongoing critical criterion of distributive justice: 'Would all people agree?' For Rawls the contract became a foundationalist principle, yet Rawls remains in part at least a consequentialist – a fact which puts him at odds with Hayek and Nozick. Rawls, however, has moved from first-level political ethics to the second level. 'Political liberalism' assumes a plurality of reasonable comprehensive moral doctrines. In what kind of reasonable society can comprehensive moral doctrines be negotiated peacefully to arrive at political judgements? For Rawls, ecological values constitute just such a comprehensive moral doctrine. Yet Rawls apparently fails to see that in today's globalised world the institutional conditions do not exist for such an entry or negotiation.

In communitarianism we find an intense pessimism and grief at the loss of the local, small-scale community with its capacity for generating a dialectic of justice. Equally we note the failure of communitarians to explore the potential for a global community, a 'community of communities', or in Rawls's terms 'a social union of social unions'. What seems to us in need of further definition is not the path back to an exclusively local scale, but new ways of inscribing the virtues of 'community' at the global level in such a way as to enable the flourishing of local dialectics of justice in local communities.

Such a global community would need to adopt an 'ethics of discourse' as part of its second-level political justice ethic, or as Karl Otto Apel puts it: 'a macroethics of co-responsibility' (1990). The second level can be approached via a politics *outside* the central spheres of governance of the administrative state, as Dryzek argues, but there remains the problem of how to control the administration of global economic governance which already exists, and how to ensure that enlarged thought penetrates the operation of this system. The increasing salience of virtual irreality in the corporate sphere distances the operation of the global economy from contact with the real roots of its existence in nature and from its real purposes in human welfare.

By the end of Chapter 4, then, we shall have explored the main dimensions of tension in the dialectic of justice: between ethics and other systems of thought, between justice and other ethical systems, among the bases of justice themselves and among different systems of political justice. We shall have identified three levels in the dialectic: moral systems (level 1), political–ethical systems which seek to resolve moral questions (level 2) and global political–ethical systems through which debates about political–ethical systems can be conducted (level 3). We touch only lightly on the relationship of this discourse to that of the environment, 'nature'. In Chapters 5 and 6, we bring the environmental question to the foreground, first by examining the distributional question and struggles (environmental justice), and then by reconsidering the interpretation of human autonomy in the light of ecophilosophy (ecological justice).

In Chapter 5 we discuss the use of conceptions of justice by geographers in evaluating human spatial–distributional systems. The coherence and explanatory power of Marxist theory provided geographers with a critical value base with which to make their work relevant for the purpose of ameliorative or transformative action without having to venture too far into the complex and unfamiliar territory of ethical discourse.

The ethical sphere has been more directly confronted by grass roots movements in the USA demanding an end to 'environmental racism' in the disposal of hazardous chemicals. The social polarisation noted by geographers has created a situation in which environmental health and safety is again problematised politically, as it was at the time of the European industrial revolution. Expanding the ethical base of the movement has entailed a definition of 'environmental justice' based on a 'rights' vision of justice appropriate to the American institutional context. Potential conflicts between the antihumanist tendencies of environmental ideologies and the humanism of the civil rights movement in defence of people of colour were largely submerged.

As industries bearing environmental hazards have been moved away from the 'developed' world, environmental justice has acquired an international dimension in the 'traffic in risk'. We consider the case of the Bhopal disaster as an instance of the insertion of hazardous industry into a less regulated environment. Awareness of the unknown aspect of risk before the event introduces new elements into the local dialectic of justice where it intersects with the global economy (with its narrowed utilitarian-entitlement ethical base). Local grass roots movements have not yet been able to mobilise at the appropriate, namely global, scale. But the dangers of the new 'risk society politics' – to use Beck's phrase – are evident. The dialectic of local environmental justice, we argue, is highly restricted unless it can also be conducted at the global level.

However, the distribution of an 'environment' for purely human use is itself too restricted to encompass an important dimension of the dialectic of justice. We must also consider our (human) moral relationship with the non-human world – a relationship of ecological justice. This relationship we explore in Chapter 6. To make sense of this relationship we re-examine the 'self-picture' which forms the ontological substrate for the bases of justice. This 'picture', we insist, is not just a figment of the imagination but a model of reality. The 'bounded self' viewed as an isolated 'ego', the normal self-picture which underpins most modern ethical systems, has been found to be deeply unsatisfying.

Ecophilosophers have sought to recover the expanded horizon of the self in three ways: by expanding the moral environment; by expanding the social environment; and by expanding the self. In the first category we include those who in different ways have tried to break down the barrier of moral considerability between humans and non-human nature – most notably by regarding animals as the subject of rights. This attempt is problematic in two ways. First, differentiation is required within non-human nature. Second, there is the real danger that once the abolition of the moral boundary between humans and animals is widely accepted, standards of treatment of humans by humans could decline to the present treatment by humans of animals, rather than the desired reverse.

Ecosocialists and ecoanarchists have found an expansion of the horizon of the self in an enlargement of the social environment to include non-human nature. Some of the founders of socialism themselves sought such an expansion. We discuss the antipathy of ecosocialists and ecoanarchists for the ecocentrism of 'deep ecology'. The former, we note, insist on anthropocentrism – humans and not Earth first. Some ecosocialists conceive of a human–animal continuum, though without suggesting useful differentiations within the continuum (but with humans at the head of the 'moral considerability' queue). Some ecofeminists find common cause with nature as an object of domination under patriarchy, without being able to suggest a stronger ethic with which to resist this domination. With the third group, the expansion of the self, we go back to the moral philosophy of Spinoza who influenced both Naess and Mathews. Naess, whose philosophy contains contradictions, finds the resolution of the question of differentiations in the human–animal continuum in political judgement, Mathews offers a strongly grounded rationale for a consistent differentiation.

25

In Chapter 7 we bring together environmental and ecological justice at level 3: global political–ethical systems. We argue that political and conceptual struggles over environmental and ecological justice can today have little practical effect on the international system which determines the fate of the planet. How, then, might this system be made permeable to the dialectic of justice? We first consider the international political economy as a system of production, and then as a system of governance. It is at this level that we choose to take a stand on principle. Since the continuation of the dialectic is central to our conception of humanity, it is axiomatic that the underlying principles of a global system must be consistent with a view of humanity in which practical reason (that is an inclusive debate having practical results) at all subordinate levels is fostered.

In the first aspect of the global system, the system of production, we discuss three perspectives: market environmentalism (ME), ecological modernisation (EM) and ecosocialism. ME theorists advocate a further major shift to the market as the solution to ecological problems. We argue that such a shift ought first to be discussed and sanctioned by a world democratic forum, rather than accomplished piecemeal and largely in secret. EM theorists have found the present direction of capitalist development highly problematical. Some versions of EM point in the direction of major institutional change but coherent guidelines for change are not offered beyond increased use of the 'negotiated order' of the United Nations system. Ecosocialists want profound changes to both the political–economic and political–institutional frameworks of globalising capitalism. But the first steps on the road to institutional change have not been specified.

As a system of governance, we find that the 'Westphalian' system is still largely intact, though modified on the one hand by the United Nations system of 'negotiated order', and, on the other, by the market-steered global economy. The latter has radically reduced the autonomy of nation states and is preventing their applying appropriate environmental regulation. Extension of the market to new environmental spheres without effective regulation will, we argue, further erode the capacity of states to protect environmental values. The negotiated order of the United Nations provides some useful models of consensual regulation of development in specific environmental fields, but it is neither comprehensive enough, nor powerful enough (lacking resources), to deliver either environmental or ecological justice.

We support the principle of transforming the system of 'world governance' which exists at present, into a system of 'world government'. We consider objections to world government but find that, with an appropriate constitution based on a reformed United Nations, a form of world government would be better able to protect both local autonomy and the planetary environment than the present system. A reformed system should not be formed as a monolithic pyramidal state, but as a nexus of accountable authorities. A first step towards a democratic and constitutional process for the formation of world government would be the creation of a (directly) democratically elected World Environment Council and an (appointed) International Court of the Environment.

World transformation in the interests of environmental and ecological justice cannot be accomplished by a rapid revolutionary process. We can only suggest first steps. Once these steps have been taken the momentum of the dialectical process of 'finding justice' can take its course with all the interests that will necessarily have to come into play. The atrophy of ethics in the world system which is preventing the dialectic of justice from expanding to include the environment and non-human nature will gradually be reversed.

CONCLUSION

The public response to the environmental crisis is moving into a new phase. From the 1960s to the 1980s the response was marked by attention to scientific studies, to the global consequences of environmental exploitation and the belief that the world could not be so irrational as to pursue a path leading to certain catastrophe. Little attention was paid to the ways in which this environmentally rational perception might be put into practice – or to whether long-term rationality might in the end be undermined by the short-term rationality of the pursuit of sectional interests. There was an underlying hope that some kind of consensus would prevail. This period saw the growth of worldwide organisations like Greenpeace as well as green parties whose aim was to keep environmental priorities on the political agenda, to create, in effect, a 'green politics'.

From the 1980s the perception grew that the environmental crisis was not only a scientific but also a deeply political matter. Furthermore, environmental politics was not going to be marked by consensus. On the contrary, there was deep division on almost every political question raised by the new environmental politics. Not only were there going to be losers and winners of environmental stakes, but there was no agreement on the basis for adjudicating such distributional matters. Optimistic formulae like 'sustainable development' began to be suspected of concealing invidious choices. More and more it appeared environmental distributions would be settled by power struggles. And these power struggles, whether they are played out on the global or national stage, always take particular local forms contingent on local cultures, local aspirations for control, and local political histories and institutions. Despite the immense economic strength of the corporations and the coercive power of states, the outcomes of these power struggles is not by any means a foregone conclusion. The profitability of corporations and the stability of political regimes remain dependent on the peaceful co-operation of local populations.

In the chapters which follow we will first explore the question of justice. We will then be in a position to discuss the ways in which environmental and ecological thought intersects with justice. Justice *is* in question. It is a contested concept. Not only is the nature of justice contested, but philosophers also question whether justice itself is an appropriate concept with which to confront the choices to be made in modern society. In the next chapter we will argue that, in the face of global environmental problems, we cannot do without ethics which are, essentially, universal. But, in the light of the cogent critique of universalism on the part of post-modernist political and social philosophers, this point has to be argued and, in fact carefully qualified. The existence of divergent rationalities appears to follow from the pluralism of thought required for human freedom.

2

DISAPPEARING JUSTICE?

'Justice' was an almost forgotten term, no longer mentioned because, like 'liberty', it had subversive overtones.

(Isabel Allende, 1988: 220)

INTRODUCTION

Justice is a complex and contested idea because it is connected with other ideas which explain society and which explain the 'self'. How, in justice, we *ought* to behave socially is connected with how we perceive that society actually *is* structured. For instance, if we perceive society to be an aggregate of individuals, we will adopt different ethical standards from those we will adopt if we perceive society to be class-structured. These perceptions are not figments of the political imagination. There is a social reality. Morality and ethics are thus connected with ontology. Moreover, how we think we *ought* to behave is connected with what sort of persons we consider ourselves to be. The basis of disagreement and contest thus extends in at least three dimensions: on what society is; what the person is; and what we ought to do. In this and the following two chapters we unravel some of this complexity. Justice and the explanation of society go together. Sometimes justice is in the background, sometimes in the foreground.

In some explanations of society justice recedes so far into the background that it almost disappears. There are two ways in which justice is made to disappear: it may be banished along with ethics in general as a useful way of thinking about society and action; and, second, justice may be replaced by other conceptions of ethics considered to be more relevant to understanding our place in the world. In this chapter we discuss explanations of society which background justice to the point of disappearance. We first consider the disappearance of ethics, and then of justice. In the first section of this chapter we examine three perspectives which put aside *ethics*: Marxist functionalism, Weberian rationalism and post-modern relativism. In the second section two perspectives in which *justice* disappears are considered: communitarian and feminist critiques of justice in modern society.

29

In the final section we discuss the relationship between justice and rationality in systems of political thought. Even though we seek to retrieve justice from the chiaroscuro of political debate (or rather from the 'oscuro'), we argue that justice has to be connected with ontologies of society and the person. Here we foreshadow the discussion in the following chapters in which we move from the bases of justice in key 'maxims' (justice in the foreground) to a fuller discussion of political justice in which rationality is combined with a justice ethic.

THE DISAPPEARANCE OF ETHICS

There are three main kinds of discourse in which either the need for ethics or the coherence of ethical discourse is rejected. Marxists have traditionally regarded ethical discourse as emerging from relations of production. In this view, ethics is part of an ideological superstructure used to justify a relationship of power. Change society and ethics will likewise change. A second view is that the primary question concerns not the ethical but the rational basis of society. This is a view associated with Weber rather than Marx. Rational steering mechanisms move societies towards certain ends. Change those ends and the principles governing the steering mechanisms will need to change. Ends are ever variable and contested and will be settled by political struggles and the exigencies of material crises. What is important is to show how different principles of organisation serve different ends. A third view is that the universalism of ethics not only reflects, but embodies power relations. Unlike Marxists, post-modernists see no solution to social problems in the 'science' of historical materialism. They oppose all universal solutions, both ethical and rational, as 'totalising' and oppressive.

The Marxist critique of ethics

There is an important passage in Wood's (1981) work that tells us much about a certain Marxist attitude to ethics:

> If we ask Marx whether the whole history of capitalism was a 'good' or a 'bad' thing, whether this history 'ought' to have occurred, then the only reply we should expect from him is a rejection of our questions as shallow, pointless and inane. Capitalism has been a terrible thing, but also a necessary thing. It has caused monstrous suffering and human waste, but it has also created unprecedented potentialities for human freedom and fulfilment. The 'good' side of capitalism cannot be separated from its 'bad' side. The important thing now is to seize on the opportunities capitalism has created, which lie beyond capitalism itself. To that end, our sympathies must lie with that historical movement which has the power to realise fully the potentialities of our

30

historical situation. This movement is, of course, the movement of the
revolutionary proletariat. Our task is to join with this movement, to
mould it and be moulded by it, to breathe the air it breathes, to let its
action be our action, its history our history.

(Wood, 1981)

In this argument it is the material reality of power relations and their historical movement and change that counts, and not the ethical ideas which
emerge at a given period of history to justify those power relations.

Yet is Wood right in his assessment of Marx's attitude to ethics? Marx's
work is infused with judgements condemning and praising both the practices
and ideas of his day. The very judgement 'capitalism has been a terrible thing'
is typical of Marx's own judgements embedded in his arguments. Marx's
critique of the exploitation of labour is constructed around an idea of fairness.
Capitalism is possible as a result of the appropriation by the employer of a
surplus (surplus value) produced by the employee for which the employee is
not compensated. Such an appropriation can only be called 'exploitation'
because it is not justified by the very norms of justice which capitalist society
claims as its own. Marx's critique is an immanent critique aimed at a crucial
flaw in capitalist society in the application of its own norms.

Marx himself subscribed to the principle, 'from each according to their
abilities, to each according to their needs', a maxim of justice. But Marx
chose to let the basis of such judgements remain inchoate. Marx saw his task
as exposing the nature of power in capitalist societies. Ethical ideas tended
not only to cloud the picture of power he wanted to present, but also to be
part of the power structure he observed. However, as Marx himself emphasised, the critique of power is conducted with a view to judgement and
action. To say, 'It is not fair' is a call to change something. It is significant
that one of the most eminent Marxist geographers, David Harvey, has
conducted his analysis of the spatial dynamics of power in capitalism always
against the background of a concern with justice. The distributive mechanism of a city is unjust because the fundamental rules and norms through
which urban development is produced are unjust (Harvey, 1973; 1992;
1993b).

When Marxists engage in political action, they immediately confront
ethical questions. For example, in a socialist society, to what degree is a
political authority justified in suppressing the political demands of those
who are not part of the 'revolutionary proletariat'? This question is neither
shallow nor pointless and it begs another question: who is *in* the 'revolutionary proletariat' and who is *outside* it?; and yet another: who is to decide
who is considered inside or outside the revolutionary proletariat? These are
the practical political questions which any potential supporter of the
revolutionary proletariat will want answered prior to committing support.
The moment these questions are brushed aside in the rush to sieze power by

violent means, in that moment the door is opened to the abuse of power. In that moment the question of the institutions of governance is left to be decided not on the basis of principle but on the basis of power and personality, on whether the leadership is of a humane disposition or is cruel, paranoid and tyrannical. There have been all too many examples of the latter, from Stalin to Pol Pot.

The need for political action has caused a number of Marxists to return to the question of how Marx's inchoate ethics might be developed. One considered interpretation of Marxism is that the Marxist critique is targeted not at ethics *per se* but at the particular historical form of 'bourgeois' ethics. According to Lukes (1985), for example, the critique of ethics mounted by Marx and Engels is a critique of 'right' or *Recht* – the bourgeois and capitalist interpretation of rights and justice underpinning the state. Now *Recht* comes into play when there are material conflicts to be resolved between persons, and especially between classes. Lukes explains Marx's scheme for understanding the bourgeois conception of justice as *Recht* which is implicated in the exploitation of labour. *Recht* supports the principles of contract and market which permit the kind of injustice involved in exploitation. The Marxist alternative was, according to Lukes, *emancipation*, that is the freeing of society from the need to resolve material conflicts by abolishing conflict at its base, namely class conflict. Ecosocialists argue that environmental conflicts over the distribution of environmental value ('good' and 'bad' environments) are rooted in class conflict, and that the source of ecological conflicts is humans' alienation from nature under capitalism (see for example Pepper, 1993).

Marx wrote that:

> nature is Man's *inorganic body* . . . Man *lives* on nature . . . nature is his *body*, with which he must remain in continuous interchange if he is not to die. That man's physical and spiritual life is linked to nature means simply that nature is linked to itself, for man is a part of nature.
>
> (Marx, *Economic and Philosophic Manuscripts of 1844*,
> [1844?] 1977: 72–3, his own emphasis)

Pepper observes that Marxists offer a dialectical view of the society–nature relationship:

> This holds, first, that there is no separation between humans and nature. They are part of each other: contradictory opposites, which means that it is impossible to define one except in relation to the other (try it!). Indeed, they are part of each other: what humans do is natural, while nature is socially produced.
>
> (Pepper, 1993: 107)

Humans and nature 'interpenetrate' one another (Parsons, 1977). There is a society–nature 'metabolic interaction' (Smith, 1984). Marx saw that capitalism would transform societies in such a way as to free them from subjection to the natural world. This was something he applauded. But he also saw that in doing so this transformation would set persons *apart* from that of which in reality they are an organic part – both socially and naturally. This separation and individuation of the human person would create an 'other', to which we ourselves are alien (see Bottomore *et al.*, 1991: 11).

Hegel, the idealist, conceived of the self as engaged in a dynamic process of alienation and de-alienation. Alienation of the self occurs *within Nature*, and de-alienation is achieved in the Finite Mind, 'man'. Marx, the materialist, sought de-alienation through the transformation of the mode of production. Communism for Marx was 'the reintegration of man, his return to himself, the supersession of man's self-alienation' (Marx cited in Bottomore *et al.*, 1991: 13).

It is surely not a distortion but an extension of Marx's thought to add that this process of alienation, and alienation of the process of production, also separates humanity from its ecosystem (other species especially) broadly conceived. This perhaps is what Marx meant when he said that alienation was man's estrangement from the nature in which he lives.

Marxists such as Nielsen (1979; 1980; 1988), Elster (1983; 1985), Heller (1974; 1987) and Peffer (1990) have constructed different versions of Marxist ethics and conceptions of justice. The general point to be drawn from these analyses is that there is no avoiding ethical questions merely on the grounds that the only concepts we have to guide us are necessarily structure and culture-bound (for Marx 'bourgeois').

Political movements, whether or not they are movements of a revolutionary proletariat, need to be able to present institutional alternatives based on realistic ethical principles if they are to command mass support. It seems reasonable to insist, moreover, that such institutional alternatives provide for the resolution of conflict in conditions of relative scarcity and moderate egoism (Hume's 'circumstances of justice'). As we have seen, some of the most pressing environmental questions are also distributional questions whose solution simply cannot await the coming of 'true communism'. Peffer (1990: 13) argues that 'only socialism – as opposed to fully fledged communism – is a practical historical possibility, at least in terms of the near and medium future'. Since socialism is characterised by both moderate scarcity and a state, he argues, Marxists need a theory of 'the right' and must confront all of the problems found in traditional social and political philosophy.

Social choice mechanisms and ecological rationalism

We do not have to use an ethical concept like justice in all circumstances of choice. In fact in the modern world, as Max Weber (1964) told us, a number of mechanisms or systems predominate which depend on simple rational rules through which choices are continually made. Weber argued that these systems are not ethical structures, they are useful (instrumental) structures.

They work to co-ordinate and channel human energies in such a way that the overall result is an improvement in the general conditions of life of the population, or so it is argued (e.g. by Hayek, 1944; Berger, 1987). They work by co-ordinating the universe of individual choices. They contain incentives which almost unnoticeably push people towards mutual service rather than mutual destruction. Though one can be excused today for thinking that 'the market' is an end in itself, its only justification is as a *means* for the pursuit of human welfare. The market is a mechanism for social choice, so also are states, legal systems and bureaucracies. Do such social-choice mechanisms make ethics redundant?

The question of the relationship between rationality, rational choice and ethics is one that has preoccupied western philosophers throughout the modern era. We shall not address that question in general terms here. What concerns us is the particular tendency for systems of rationality to displace ethical discussion. Let us accept, for the moment, Berger's proposition that, 'An economy oriented toward production for market exchange provides the optimal conditions for long-lasting and ever-expanding productive capacity based on modern technology' (Berger, 1987: 37). Berger's view of capitalism as a 'horn of plenty' is based on the uncritical Promethean assumption that nature is an infinite bounty (*ibid.*: 35). What is today in question is whether 'ever-expanding productive capacity' can be consistent with preserving the ecological integrity of the planet. But suppose that a social-choice mechanism could be found, or existing mechanisms adapted, so as to co-ordinate individual choices in such a way as to guarantee ecological integrity. Would such a mechanism make it unnecessary, or at least less important, to find an environmental ethic? Can a rationality without ethics be the foundation for environmentalism?

Dryzek (1987) has sought to specify the conditions which social-choice mechanisms would have to fulfill if they are to give rise to the preservation of ecological integrity. He argues cogently that ecological problems are today so serious and so threatening to the survival of humanity that a radical reconstruction of the basis of social-choice mechanisms must now be attempted. Dryzek, it should be noted, does not regard the market *a priori* as a 'better' mechanism than, say, democratic states, for making social choices. In fact he recognises that most societies combine a variety of mechanisms in different ways (hierarchies of command, markets, bargaining, anarchy, voting, legal systems, persuasion, and even armed conflict). Dryzek's contention is that, 'the nature of the collective choice mechanisms in place will largely determine the kind of world that ensues' (*ibid.*: 8).

The point of Dryzek's analysis is to develop a set of criteria against which to evaluate existing social-choice mechanisms. The criteria constitute an instrumental value set. They are values derived from the sort of problem which ecological problems are: how such a problem has to be addressed in order to meet certain ends. The ends to be pursued are not the primary focus of attention. Dryzek deliberately restricts his consideration of ends to the anthropocentric: the capacity of the earth's ecologies to continue to support human life; the need for continuing production for human use, the stabilisation of the human ambient environment through the 'buffering of air and water cycles, the moderation of temperature extremes, and the regulation of the abiotic environment (for example, atmospheric ozone and physical substrata' (*ibid.*: 34–5), and the capacity to absorb human waste products.

Dryzek agrees that such a value set is minimal and does not in any way exclude or diminish other ways of valuing the environment. It is adopted because it enables him to confront other anthropocentric rationalities on their own grounds. Within examinations of social choice there is an inchoate ethical debate. The ethical question is not denied but it is passed over fairly quickly in order to focus attention on the means. But ethical issues have to be confronted, for the reasons discussed above in connection with certain Marxist denials.

One observation about the current treatment of environmental problems to which Dryzek devotes much attention should alert us immediately to the importance of justice as an ethical issue: the tendency of 'displacement'. Dryzek uses the problem of acid rain as the paradigm of an ecological problem. The problem may be 'solved' by shifting the impact of its effect. Thus the 'problem' of sulphur dioxide emissions is 'solved' by building tall smokestacks: 'Instead of polluting areas adjacent to copper smelters in Utah or coal-burning power stations in Ohio, the sulfur dioxide ends up in the form of acid rain in rural areas such as the Rocky Mountains or the Adirondacks' (*ibid.*: 16).

Dryzek notes three forms of displacement. First, the problem may be displaced in space. Toxic waste is shifted from one dump to another or from one country to another. In more general terms, polluting and environmentally damaging industry is shifted to countries with weak environmental standards. From the example given in Chapter 1, mining at Ok Tedi causes a displacement of the environmental costs of the use of raw materials to a country which is prepared to trade environment for 'development'. Second, an environmental problem may be 'solved' by displacing the problem to another medium. The problem of disposal of the oil rig, Brent Spar, was 'solved' by reversing the decision to dump the rig in the sea and displacing the problem to be dealt with on land and in the air. Third, a problem may be solved by displacing it into the future. French nuclear testing in the Pacific will almost certainly create a future problem of radiation leakage, though how far in the future the problem will become manifest can only be guessed at.

Whether or not the ecology of the planet as a whole becomes overloaded by the aggregate impact of environmental problems, there is a continuing question of the distribution of impacts. The environmental justice movement in the USA arises precisely from one form of spatial displacement: the shifting of the burden of toxic waste disposal into neighbourhoods occupied by people of colour. Harvey (1981), Smith (1984) and others have argued that displacement, an aspect of 'uneven development', is not some unfortunate aberration but fundamental to the dynamics of capitalist development. Dryzek's discussion of ecological rationality, as he himself has articulated in later work (1990; 1994; 1996), does not make ethics redundant.

Dryzek claims that ecological rationality should have 'lexical priority' over economic, social, legal or political rationality, 'lower values come into play only when designs in pursuit of a higher value are totally complete' (Dryzek, 1987: 59; and cf. Rawls, 1971). The primacy of ecological rationality rests, in Dryzek's view, on the highest (most basic) value: survival. But rationalities of any sort do not and should not do away with politics. And while there is politics – involving unequal power and conflict – the question of justice is central. Dryzek (1990; 1994) recognises that in order to solve ecological problems, new political institutions have to be created, and, what is more, the *means* for creating new institutions. These institutions and means have to measure up to ethical standards. Unfortunately, however, if we relegate justice to the 'second division' of ethics, 'ecological rationality', sustained by the 'primary value' of survival, may be used to justify ecototalitarianism (see Ophuls, 1973; also the warnings of Ferry, [1992] 1995 and Bookchin, 1995b).

Justice is a value which inheres in political or legal rationality, secondary rationalities in Dryzek's view. One can imagine competing claims being made that a particular political/economic system is necessitated by the primary value of survival, and which thus takes precedence over any alternative. One can also imagine ecological rationality becoming the basis of a totalising project justifying authoritarianism. Ferry's ([1992] 1995) discussion of 'nazi ecology' and Bookchin's (1995b) warnings of ecofacism should give us pause for thought. In a reasonable political society, what is 'primary' and what is 'secondary' must surely be considered fair territory for political struggle and discursive debate.

Eventually, a discussion of ethics will end up at the same sort of destination as discussions which eschew, or at least do not focus upon, ethics. That destination is the construction of institutional systems, perhaps even global systems, for handling power and conflict in human society. The questions such institutions will have to confront are ethical questions: what is justice in this or that context? Such systems do not dispose of ethics but must themselves have an ethical basis.

Post-modernism and value relativism

'Listen', says Thrasymachus in Plato's *Republic*, 'I define justice or right as what is in the interest of the stronger party'. The argument of Thrasymachus against Socrates is echoed by Bauman when he writes:

> Whatever is agreed 'truly universal' in the end, is more in the nature of a 'common denominator', rather than the 'common roots'. Behind the

procedure lurks the assumption that makes it workable: that there is more than one conception of universal morality, and that which one of them prevails is relative to the strength of the powers that claim and hold the right to articulate it.

(Bauman, 1993: 42)

This is a version of the Marxist critique of ethics. Again it is power that counts. Ethics follows and entrenches power. But post-modernists reject even the possibility of a universal ethic and attack the claim that there is one correct way of viewing society – whether that way is as a free market or an exploitative class structure. Lyotard (1984), for instance, does not merely make claims about how social systems ought to be organised but about how we come to know the world, and thus how we come to know society. Comprehensive theories of both the world of objects and the world of persons are seen as finite instruments working within certain standards of knowledge, or 'narratives' which legitimate knowledge and define truth.

In both science and social theory, Lyotard claims, we can now recognise not one dominant narrative or even two opposing narratives but many. The 'grand narratives' raised in opposition to each other have broken down: 'the grand narrative has lost its credibility, regardless of what mode of unification it uses, regardless of whether it is a speculative narrative or a narrative of emancipation' (*ibid.*: 38). There is no 'text beyond the text'. There are many versions of the truth with no way of choosing among them other than by means of purely instrumental criteria. In criticising Habermas for his 'modernist' insistence on the necessity of consensus at *some* level, Lyotard writes, 'Consensus has become an outmoded and suspect value. But justice as a value is neither outmoded nor suspect. We must thus arrive at an idea and practice of justice that is not linked to that of consensus' (*ibid.*: 66).

This may well be so, but Lyotard does not provide us with any insights into how justice might be reinscribed in the fragmented but globalised world of networks he portrays. Arran Gare (1995) has examined the contribution of the post-structuralists to environmental issues from a mostly sympathetic point of view, commending the insights of Derrida, Lacan, Barthes, Baudrillard, Lyotard, Deleuze and Guattari. In the end, however, he finds them wanting:

With their opposition to extra-texts beyond the text and to texts which would sum up other texts, post-structuralists leave environmentalists no way to defend their belief that there is a global crisis or to work out what kind of response is required to meet it. . . . The way they have developed their scepticism about any simple relation between language (or texts) and 'reality' has engendered, or at least supported, a form of idealism in which intellectual life is increasingly centred on the discussion of what other people have said.

(Gare, 1995: 98–9)

Post-structuralist critique correctly recognises power in discourse. An ethic, a morality which claims to be universal, is a structure of power. But there would be little point in such a critique without values of some kind. Otherwise, as Soper (1991: 123–4) observes, 'why bother?' What seems to substitute for values in much of the post-structuralist narrative is the myth of a world without power. Bauman (1993: 45), however, continually returns to the task of constructing a universal ethic, indeed a global ethic. He writes:

> Moral responsibility prompts us to care that our children are fed, clad and shod; it cannot offer us much practical advice, however, when faced with a depleted, desiccated and overheated planet which our children and the children of our children will inherit and will have to inhabit in the direct or oblique result of our present collective unconcern.
>
> (Bauman, 1993: 218)

For Bauman 'the moral' must be defined in the most demanding form possible, namely: 'Do to others what is best for them'. We have to face up to the extreme difficulty of carrying out such an imperative. If we do face up squarely to this difficulty, we will see that truly moral action has to be *negotiated* because, of course, we can very rarely know what is 'good for them'. Finding 'the good' is a continous struggle in which the possibility that we may be wrong is ever present. A global ethic, certainly an environmental ethic, also has to be continuously negotiated. There can be no once-and-for-all-time rule. The problem is to create a public politics at the global level at which moral questions and ethical judgements can be wrestled with and not smothered by preconceived rules. Yet, if we must face the extreme difficulty of making moral choices, must we not also face the fact that rules of some sort are indispensible in the creation of the institutions of a moral politics, and that those rules must apply to the whole population of our planet?

The acknowledgement of difference need not lead to the end of transcendence or to social paralysis but to a further exploration of the rules proper to our humanity for the resolution of conflict and the regulation of action. Acknowledging the validity of different cultural norms might actually facilitate an intercultural dialogue which provides a route back to universal principles. The reassessment of the importance of the particular, of the context of a *judgement*, need not lead to a reduction in the importance of the universal, of *justice*. Thus, we are led towards a reassessment of the role and function of universal principles in acts of judgement. If we are to make judgements about such matters as the sinking of oil rigs, the mining of rainforests and the use of islands for nuclear testing, then we require a 'political ethic' which 'concerns the creation of institutions, the formation of practices, and the sustaining of civic values that cultivate the ability of enlarged thought and the universalist–egalitarian commitment which inspires them' (Benhabib, 1992: 139).

THE DISAPPEARANCE OF JUSTICE

If ethics are indispensible to a discussion of political issues, not least those concerning the environment, there remains the question of whether a 'justice ethic' is appropriate to the relationships involving people, their environments and nature. There are two major critical discourses within which the ethic of justice tends to disappear. The first is the communitarian critique of liberalism, the second a certain strand of feminist thought.

The communitarian critique of liberalism

The idea of creating local communities equipped with values sensitive to ecological sustainability and supporting just relationships between human and non-human aspects of nature is an important and recurrent theme in the literature and politics of the environment. Nisbet (1974) described the human ecological community with the capacity to live in balance and harmony with the land and the non-human natural world. Roszak's (1978) idea of 'ecomonasticism' provides a model of the stable domestic community which offers 'egalitarian fellowship' and satisfies the demands of justice to other humans and nature. Sale (1980; 1985) extends the idea of small self-sufficient communities to regional confederations defined by reference to biotic communities. Bookchin (1982; 1990) erects the idea of the decentralised human-scale community in opposition to the modern hierarchies of nation states. LeGuin (1986) writes lyrically of the life of a self-sufficient community in the aftermath of an unspecified ecological catastrophe. Such ideas have a long history as Eckersley (1992: 160–70) and Pepper (1996: 210–17) elucidate. Should we then concentrate on rediscovering or reformulating the 'ecological community' and leave the discussion of justice to the communities themselves?

In one of the strongest statements of the communitarian position, Alasdair MacIntyre (1981) argues that, although today we continue to talk about morality and ethics, we have lost the political tradition which gave meaning and practical effect to the very precepts we use in our ethical language. So we are left with fragmentary notions which we can only piece together in a kind of tattered patchwork without any underlying coherent structure which would tie together concepts, norms and practical conduct. Our concepts of justice, Macintyre claims, are incommensurable because of the absence of the sort of community in which what is just and unjust becomes defined.

The society we have today is one in which people behave to one another as 'nothing but a collection of strangers, each pursuing his or her own interests under minimal constraints' (*ibid.*: 251). In a world in which our desires are defined without reference to the good of anyone else, in which the good of another cannot be part of our conception of the good of our 'selves', in which 'the individual' is to be treated as an isolated atom, in such a world the only possible conception of justice is one in which these isolates and strangers enter into contracts to secure their own (limited) good. Macintyre argues that the plurality of our concepts of justice has a certain function in our individualistic society.

> It is not that we live too much by a variety and multiplicity of frag-mented concepts; it is that these are used at one and the same time to express rival and incompatible social ideals and policies and to furnish us with a pluralist political rhetoric whose function is to conceal the depth of our own conflicts.
>
> (*ibid.*: 253)

Sandel (1984: 161) argues that the abstract assumptions of liberalism do not correspond well even with the reality of modern society which is supposed to embody them. He takes issue particularly with two assumptions of liberalism: the radical separation of thinking and feeling and the claim that it is impossible for one person to know what is for the benefit of another. The latter, in particular, Sandel considers, is a convenient way of avoiding moral responsibility. We are subjects, he posits:

> constituted by our central aspirations and attachments, always open, indeed vulnerable, to growth and transformation in the light of revised self-understandings. And in so far as our self-understandings compre-hend a wider subject than the individual alone, whether a family or tribe or city or class or nation or people, to this extent they define a community in the constitutive sense.
>
> (*ibid.*: 166)

Others (for example, Taylor, 1979; Bellah *et al.*, 1985) have argued that the liberalism of capitalist societies requires the virtue of actually existing communality in order to function at all. As we will see later (Chapters 4 and 6), conceptions of the self underlie all political ethics.

Taylor, writing of Hegel's relevance to modern politics, discusses Hegel's distinction between *Moralität* and *Sittlichkeit*. The latter 'refers to the moral obligations I have to an ongoing community of which I am part' (Taylor, 1984: 177). *Sittlichkeit* always emerges from a particular community which already exists. *Moralität*, on the other hand, is the abstract, the universal, the ideal which is used to bring about a state of affairs which does not exist. Taylor shows us how Hegel, at a time when capitalism was beginning its social dominance, wanted to recover the kind of community in which ethical behaviour would again become part of public life. Capitalism created the individual cut off from society, alienated, isolated. Hegel applied his *Moralität* to define the organic community, with the state at its core, which would generate ethical behaviour. Today, in the light of our understanding of global society and its consequences, we surely have to return to the task Hegel began.

Immanent in communitarian thinking, from Hegel on, is the possibility of changing mass society itself, to reconstruct a form of community capable of articulating conceptions of justice. In MacIntyre's sense of loss there is the unexplored possibility of retrieval. In Sandel's conception of the wider 'self' there is the possibility of 'a people' becoming constituted as a community. In Bookchin's 'community of communities' there is the intimation of a global community. But how to begin achieving these possibilities at global scale has been little considered by communitarians.

It is difficult to know how to define the value of 'community' separately from the value of justice. Ideas of community seem to be inseparably interwoven with ideas of justice. What we believe we owe to each other, and what our duties are to each other within 'community' (ideas of justice) are definitive of what 'community' means. So the value of community does not override justice but rather is constitutive of justice. However, in communitarian thought there is an assumption about what a community is like, namely that a community is a socially and spatially bounded set of relationships. Different and putatively incommensurate ideas of justice emerge from the cultures constituted by bounded communities. The discourse of justice which emerges from such bounded social realities is counterposed to the discourse of justice which emerges from the unbounded (both socially and spatially) relationships of modern society, the *Gemeinschaft* as against *Gesellschaft* of Tönnies.

The discourse of justice in modernity is indeed couched in abstract and universal terms. Yet, this discourse is not divorced from social reality. It is just that the social reality from which the discourse emerges is not the social reality of bounded sets of relationships. Modern thinking about justice has had to deal with the reality that increasingly the lives of persons are embedded in relationships of which the principal feature is membership of the human species. As Kant observed at the end of the eighteenth century: 'The peoples of the earth have thus entered in varying degrees into a universal community, and it has developed to the point where a violation of rights in one part of the world is felt everywhere' (*Perpetual Peace* in Reiss, 1970: 107–8); and 'Nature has enclosed them [nations] all together within determinate limits by the spherical shape of the place they live in, a *globus terraqueus*' (Kant, 1991: 158 [*The Metaphysics of Morals* III §62]). The boundaries are those of the biosphere.

The idea that cultures and communities are 'separate, bounded and internally uniform' is itself a product of early modern encounters with cultural differences (Tully, 1995: 10). Tully argues that this conception of bounded cultures is an idea which developed with modern constitutionalism. However:

41

over the last forty years this billiard-ball conception of cultures, nations and societies has undergone a long and difficult criticism in the discipline of anthropology . . . it has gradually been replaced by the view of cultures as overlapping, interactive and internally negotiated.

(*ibid.*)

If this view is correct, what has to be discovered is a conception of justice not for a homogeneous world of cultural uniformity, nor for a world of sealed-off bounded communities, but for a web of interacting, overlapping cultural diversity in which both the reality of belonging to a local community *and* to an unbounded – ultimately global – community are given due recognition.

It may well be true that ideas about ecological and environmental justice will take shape in small self-sufficient communities. But what of the society limited only by the gravity of *globus terraqueus*? Short of a catastrophe of unparalleled violence, as implied by LeGuin (1986) in *Always Coming Home*, there seems no good reason to suppose that small, largely unrelated, self-sufficient communities will replace cosmopolitan society. More likely, if anything, such communities will come and go within the larger society. The unbounded society will continue to exist whether bounded ecological communities are created or not. And for that society there is also a need to address the question of justice – which is also the question of creating it as a community.

Feminism, the ethic of care, and the principle of inclusion

Feminists have been among the strongest critics of the consumerism and productivism which, they claim, have brought the world to the current point of ecological crisis (see, for example, Mies and Shiva, 1993; Plumwood, 1993; Merchant, 1992; 1996). King demands that the domination of nature and the domination of persons be addressed simultaneously:

The domination of nature is inextricably bound up with the domination of persons. . . . There is no point in liberating people if the planet cannot sustain their liberated lives, or in saving the planet by disregarding the preciousness of human existence, not only to ourselves but to the rest of life on earth.

(King, 1990)

Women have been at the forefront of the struggle against the injustice of toxified local environments and the degradation of the planetary ecology (for example, Sontheimer, *c.* 1991; Seager, 1993; Rocheleau *et al.*, 1996).

It may seem, then, that feminists are taking up anew the struggle against injustice. But some strands of feminist philosophy engage in a substantial critique of universalism and justice. Each of these categories is a construct, it is argued, of the male-dominated culture.

With regard to the first of these, universalism, Young argues that:

> The ideal of impartiality is an idealist fiction. It is impossible to adopt an unsituated point of view, and if a point of view is situated, then it cannot be universal, it cannot stand apart from and understand all points of view.
>
> (Young, 1990: 104)

Following the Marxist tradition, Young substitutes injustice for justice: 'faces of oppression' (1990), namely *exploitation* at work and in the home, *marginalisation* – the exclusion of certain categories of persons from useful participation in society, *powerlessness* – the exclusion of certain categories from political power, *cultural imperialism*, – the exclusion of minority cultural norms, and *violence*. Pulido (1994) has used Young's framework to argue that the concept of environmental justice in America springs from a growing consciousness at grass roots level of environmental *injustice*.

What is required is an inclusive politics which preserves and encourages the cultural differences within groups, but gives recognition to the affinities and overlapping experiences between groups, for, 'different groups are always similar in some respects, and always potentially share some attributes, experiences and goals' (Young, 1990: 171). Young is against an exclusionary 'politics of difference' which demands conformity and the shibboleth demarcating insiders from outsiders. Mies and Shiva (1993: 129) comment on just such an exclusionary politics in the Balkans war of the 1990s.

Nevertheless, embedded in Young's conceptual framework, we find universal principles, a political ethic. In Young's view a political ethic must include respect for cultural differences, non-exclusivity, 'social equality', 'the full participation and inclusion of everyone in a society's institutions, and the socially supported substantive opportunity for all to develop and exercise their capacities and realise their choices' (Young, 1990: 173). A just political system, she argues, must involve a more proactive stance than the *tolerance* of difference in traditional pluralism. True pluralism requires a stance of 'listening', a stance which does not approach 'otherness' with a view to changing it but of understanding it and fostering it, 'an emancipatory politics that affirms group difference' (*ibid.*: 157). This politics depends upon a conception of 'the equal moral worth of all persons, and thus the right of all to participate and be included in all institutions and positions of power and privilege' (*ibid.*: 159).

These principles are surely not far from the principles enunciated by Kant ([c.1798] 1991) in 'The Doctrine of Right' in *Metaphysics of Morals*, and repeated by Rawls (1971): 'fair equality of opportunity'. Without such principles it is impossible for feminists to support with moral arguments their struggle against the oppression of women in patriarchal societies and groups. If these principles are *not* to be understood as universal principles, then Young must be prepared to accept the morality of societies and groups which do not accept the principles. Without some universal principles there can be no requirement, even, that such societies and groups respect the principles of others.

We dispute Young's claim that 'difference' is always to be 'fostered' and encouraged. There is certainly a case for fostering difference in some cases. But in others there is a case for merely tolerating difference. When the moderate oppression of women (as we see it) is embedded in the culture of a group, should this culture be 'fostered'? Is there not rather a case for 'toleration'? Toleration, in this instance, might take the form of attempts to bring about change through discursive persuasion rather than by means of any form of sanction or force. Intercultural dialogue requires tolerance and listening, but not necessarily 'fostering'.

Moreover, we would argue, true dialogue requires argument that one position is *better* than another. The only reasonable basis for such a claim that one position is better than another would seem to be that one position better expresses the humanness of humans, the transcendent and universal qualities we share between cultures. It will not do to say: 'Well what you say is right for you, but not for me'. Simply agreeing to differ in this way is merely an avoidance of dialogue. We would be saying in effect: 'Let us talk about many things but let us avoid talking about what centrally divides us'. In such an artificial and vacuuous interaction no-one learns from anyone and the status quo is forever preserved.

As to the second, justice, Gilligan (1982) has identified the concept of justice with a specifically masculine psychological development and a masculine gendered view of the relationship between self and others. The justice ethic, Gilligan claims, arises from the masculine drive to establish the self as a separate and autonomous entity. Different virtues are stressed if driven by the need for (or a feeling for) connectedness rather than individuation, a need most often borne in modern industrial society by women. In Gilligan's words:

> The values of justice and autonomy, presupposed in current theories of human growth and incorporated into definitions of morality and self, imply a view of the individual as separate and of relationships as either hierarchical or contractual, bound by the alternatives of constraint and co-operation. In contrast, the values of care and connection, salient in women's thinking, imply a view of self and other as interdependent, and of relationships as networks created and sustained by attention and response.
>
> (Gilligan *et al.*, 1988: 8)

Gilligan goes on to elaborate an ethic based on connection and care, rather than individuation and obligation. The contrast between the two moralities has been further delineated by Lyons (1988: 35). In a morality of justice, Lyons argues, moral problems are generally construed as decisions between conflicting claims of self and others (including society). They are resolved by invoking impartial rules considering: (i) one's role-related obligations, duty, or commitments; or (ii) standards which include reciprocity: 'how one should treat another considering how one would like to be treated if in their place'. In a morality of care, by contrast, moral problems are generally construed as issues of relationships or of response, that is how to respond to others in their own particular terms (recall Bauman, 1993). Decisions are made through the activity of care, considering: (i) the priority of maintaining relationships; or (ii) promoting the welfare of others, preventing their harm, relieving their burdens, hurt or suffering (*ibid.*).

The morality of care deserves to be considered in its own right irrespective of whether it is specifically a feminine morality. Both the morality of care and the morality of justice depend on an axiom concerning what the human person is. The morality of care postulates that the human person is connected with others. This postulate is not opposed to a Kantian morality of justice, but a Kantian would derive justice from the fact that the 'others' with whom one is connected are ends in themselves. In a morality of care 'connectedness' can be interpreted in a literal sense as being connected to those others in our immediate circle. We therefore have diminishing responsibilities to more distant others (this appears to be Wenz's (1988) view). This leads to the position that our duties and responsibilities are affected by the degree to which we can keep others who might need our care at a distance. If we can segregate ourselves from the poor, for example, we may think that we have reduced obligations to them.

This approach seems to us to confuse two kinds of response – the response of empathy and the response of justice. It is certainly true that we are liable to *feel* less connected in responsibility to distant others than to those close to us. But that diminishing empathic feeling is a separate matter from how we ought to treat distant others on account of their rights. *Caritas*, care, can become optional 'charity'. Whereas if distant others have rights then, in justice, we *must* respect them (we return to this issue in the next chapter).

However, Gilligan's own approach leads to the conclusion that the ethic of justice and that of care are complementary rather than in opposition, though the latter 'voice' has become submerged in a patriarchal society. Gilligan's argument seeks to *articulate* the conceptual basis of women's moral judgement rather than to *oppose* the ethic of justice. In the light of her concepts it becomes possible to perceive a moral coherence about some common practices in modern society concerned with human welfare based on need. Such practices can be found at the heart of the legal system itself. If the application of universal rules were all that society demanded, judges would be redundant. Good judges, in sentencing, consider the specific situations of offenders. This practice, or so one would hope, is based on an ethic of care. On the other hand, we are obliged to pay taxes to governments to help the disadvantaged in our society, because they have rights, not because we feel empathy.

Feminists have articulated a particular philosophical principle which unites their work – the principle of inclusion. What we learn from feminist scholars is not the exclusion of justice but the inclusion alongside justice of other principles, other ethics. The ethic of care does not exclude that of justice, and certainly the ethic of justice does not exclude that of care.

RATIONALITY AND JUSTICE

We have sought to retrieve justice within different perspectives in which it seems to disappear. But there are important lessons to be learned from these perspectives. First, the idea of justice is incomplete without considering its social and political context. Justice is not an abstract object waiting to be discovered. Justice is a quality of human conduct. Human conduct involves both reflection on ourselves and our being in the world, and conduct co-ordinated with others, social conduct, political conduct. If we are to discuss the justice of our distribution of environmental quality and the justice of our relationship with the non-human world, we must repeatedly come back to the 'self' in society.

Conceptions of justice in modern western liberal philosophy rest on an idea of the human person. The conventional picture is roughly as follows. The human person is an integer, whole and complete in itself with clearly defined boundaries which separate it from others. The human person is an 'atom', an 'individual', both words meaning an entity which cannot be taken apart. Humans may seek to distinguish themselves from one another, but at some fundamental level they are alike and equal. Human persons can choose their way of living. In this sense they are free. All humans are alike and *different* from all of the non-human world. This difference is a hierarchical difference. Humans are *superior* to the non-human world. Superiority is manifested in the different standards of conduct which humans apply to other humans as opposed to the moral standards they apply to the treatment of the non-human world. Moreover, one may not treat animals as things. There are thus two moral barriers: between humans and animals, and between animals and things (inanimate objects).

We will later see that the atomistic individuality of the human person has been challenged, the sharp moral distinction between the human and non-human world has been blurred, the superiority of the human species over the rest of the natural world has been attacked (see Chapter 6). But the conventional world picture has an important consequence which those who challenge it do not at all want to do away with: the fundamental equal worth of human persons. The human egalitarianism which goes with modernity leads to the view that one person's choice of carefully considered and informed ethical perspective (the opinion of an intelligent mind) is as good

as anyone else's. This 'egalitarian plateau', Dworkin (1983), argues is fundamental to justice. That is to say that, in arriving at justice, we have to take
account of the fact that people have and are entitled to hold different
perspectives.

Does this perception then land us in a situation in which no judgements
can be made because there are no universally agreed bases for judgement?
The problem recedes if we focus on human conduct, both individual and
social. Seyla Benhabib (1992), with the help of Hannah Arendt's work,
makes an important distinction between justice and judgement. She argues
that morality cannot be derived exclusively from abstract and general moral
laws or imperatives. Rather, doing justice requires the making of moral
judgements about particular situations.

These are situations in which persons interact and in which actions are
liable to be interpreted in different ways. In other words, in matters
requiring moral judgement there is almost always a moral dilemma.
Different moral rules commonly point in different directions. She says that
we have first to recognise a situation as morally relevant, that is one
requiring moral judgement. We cannot rely on universal rules to make such
an identification. We then have to exercise moral imagination in deciding
how to act. That is we have to consider what would be right in *this* case,
with *these* people.

Defining an act as moral means taking account of the fact that morality is
always intersubjectively determined. She writes, 'The self is not only an I but
a me, one that is perceived by others, interpreted and judged by others. The
perspectives of the I and the me must somehow be integrated to succeed in
making our actions communicable' (Benhabib, 1992: 129). In other words we
learn a conception of self from the judgements others make about us.

She then turns her attention to the conditions under which moral imagination can be exercised. With Hannah Arendt she argues that morality
disappears from political life when the capacity to think about action in
moral terms is removed or reduced. So Benhabib argues in favour of political
institutions in which, following Arendt, 'enlarged thought' becomes
possible. These political conditions must be such as to allow many moral
perspectives to be brought to bear upon judgements leading to action.

> The more human perspectives we can bring to bear upon our under-
> standing of a situation, all the more likely are we to recognise its moral
> relevance or salience. The more perspectives we are able to make
> present to ourselves, all the more are we likely to appreciate the
> possible act-descriptions through which others will identify our deeds.
> Finally, the more we are able to think from the perspectives of others,
> all the more can we make vivid to ourselves the narrative histories of
> others involved. Moral judgement, whatever other cognitive abilities it
> may entail, certainly must involve the ability for 'enlarged thought', or

the ability to make up my mind 'in an anticipated communication with others with whom I know I must finally come to some agreement'.

(Benhabib, 1992: 137, citing Arendt)

The introduction of new conceptions of justice based on an ecological world view further enlarges the possibilities of thought and thus of divergent and conflicting perspectives. These perspectives can be integrated in acts of judgement. These acts of judgement, if they are to be brought to bear on cases like those discussed in Chapter 1 require certain institutional structures which are themselves based upon principles of justice. The way out of relativism is to focus on the institutional conditions in which the egalitarian view of the human person can be most fully realised, an egalitarianism which inscribes an open-endedness, inclusiveness and inconclusiveness within our views of justice. Such institutional thinking directs us to combine ethical ideas about justice with ontological ideas about persons and society.

We shall have to move between principles of justice and the institutional conditions for judgement, between bases of justice and political justice. Justice to humans and nature will continue to be contested. The contest is largely, though not entirely, a human contest and the contest is itself supported by the axiom of human equality.

CONCLUSION

We have seen in this chapter that there is no escaping from ethical dilemmas. Philosophies in which ethics seem to disappear merely reveal ethics in a different guise. Where justice fades into the background, there is often a need to rediscover it. A world in which there is power and conflict must be a world in which there is a place for politics. Where there is politics, we must have conceptions of justice.

Today we are becoming more and more aware that the politics of the twenty-first century will be an environmental politics. The spectre stalking the world is the destruction of environments and planetary catastrophe. What is taking place today is not the utter rejection of universal values and the acceptance of relativism, but the rearticulation of the relationship between ethical practices and principles, between the particular and the universal, between judgement and justice. Divergent conceptions of ethics and justice are a fact of human existence. The question is how can they be reconciled, not in universal laws but in practical cases of judgement in such a way as to preserve humanity and the rest of nature. In the next chapter we turn to the plurality of conceptions of justice.

3

BASES OF JUSTICE

If we in Asia want to speak credibly of Asian values, we must also be prepared to champion those ideals which are universal. . . . It is altogether shameful, if ingenious, to cite Asian values as an excuse for autocratic practices and denial of basic rights and civil liberties.

(Datuk Seri Anwar Ibrahim, Deputy Prime Minister of Malaysia: cited in the Editorial, *The Age* , Melbourne, 11 November 1996, p. A14)

INTRODUCTION

While a 'justice ethic' cannot stand alone and should be augmented with other ethical principles, justice is an indispensable requirement for the ethical resolution of conflict *among* humans over nature, and *between* humans and nature. We now address the foundations of a justice ethic. In this chapter we shall for the most part be looking back at debates about the nature of justice and trying to understand the discourse. The primary focus is the relation of the self to human others. The issue of the relation of the self to non-human others has only recently begun to emerge. So in this chapter we will touch only lightly on the extension of the question of justice to include this relation. We shall, however, try to show how such an extension need not mark a radical break in the ongoing dialectical process of finding justice.

The analysis starts (following Miller, 1976) at the level of the philosophical 'bases' of justice. We can capture the essence of these bases in three maxims: justice is giving and getting what is *deserved*, justice lies in *rights* being respected, and justice requires that the *needs* of each be met through the contribution of each according to their ability. Reversing Miller's order of consideration, we deal first with the seemingly more culturally embedded 'desert'. Miller claims that these bases are incommensurable definitions of the meaning of justice. We shall see however that these meanings overlap, and several writers have endeavoured to show how there is a single underlying reality to which they refer (e.g. Turner, 1993; Davy, 1996). Following discussion of the 'bases' we reconsider the illustrative cases outlined in

Chapter 1. Questions remain as to whether these 'bases' of justice must now be revised in the light of new understandings of environmental ethics.

JUSTICE AND DESERT

Signalling that 'ecological justice as desert' might not be an absurdity, Feinberg (1970: 55) begins an essay on 'personal desert' by observing, 'Many kinds of things other than persons are commonly said to be deserving'. He cites 'art objects' which deserve to be admired, and 'bills of legislation' which deserve to be passed. Both are human creations, but it makes perfect sense to say that animals deserve to be treated humanely, or that the rain-forest deserves to be protected.

Feinberg (*ibid*.: 62) classifies some of the occasions on which a social action commonly involves something being said to be 'deserved'. A prize or honour is awarded because someone deserves it. Students' essays are given grades corresponding to what the merit of their papers deserve. A criminal is said to deserve punishment. We blame and praise people in accordance with what they deserve. Compensation or reparation is meted out to those who deserve it for having suffered unjustly. We could probably find many other examples. A particularly important example in the modern world is when we say, 'The government deserved to lose the election': political desert. What they all imply is that judgements are made by some people about others, and sometimes decisions are made to allocate to those people certain honours or punishments. Often, but not exclusively, the judgements and allocations are made to persons on behalf of some community or society. The paradigmatic case of desert seems to run along these lines. A person (or persons) arouses feelings of admiration or disapproval in other people. If these others are organised socially, the society will express these feelings in decisions to allocate something to the person who aroused the feelings. Whether or not the thing allocated has monetary value is irrelevant to the moral standard. But, as Hobbes (1929, Chapter 10) was quick to recognise, public honour confers power.

It would seem then that the idea of desert arises out of the customs of particular societies and cultures. In modern western societies we no longer hand out wreaths of laurel or oak leaves but we still award medals for valour and a vast range of prizes for merit. In some societies certain criminals 'deserve' to be killed or to have their hands cut off, in others the same sort of criminals 'deserve' a prison sentence and rehabilitation. It seems difficult to detach desert from its cultural and political setting. Yet to say, 'one who commits a crime deserves punishment' is a different sort of statement from, 'one who commits the crime of stealing deserves imprisonment'. We can see that the general idea that persons who do something bad deserve something bad to happen to them is at least a higher order, if not a universal, idea (retribution). It is to be found in most cultures. Yet, the definition of what is bad and good, and also *how* bad and *how* good, and what other conceptions of justice are to be brought to bear in judgement (e.g. compassion, or the consequentialist idea of 'the protection of society') can only be determined within cultures. Ferry (1995), for example, tells us that in mediaeval Europe animal pests such as leeches were thought to deserve a fair trial. There is no reason why a wide range of 'desert' claims should not emerge from societies of the future in which moral considerability is extended to the non-human world.

Does this mean that 'desert' is an idea which only emerges from the standards of particular cultures? As we have argued in the last chapter, oppression cannot be *justified* (though it might be tolerated) on the grounds that cultural standards differ and must be respected. If such words as 'oppression', 'exploitation' and 'alienation' are to mean anything at all, they must be intended to apply to all people everywhere at all times. Bauman's (1995: 182) relativistic claim that the deeds of the leaders of Nazi Germany who ordered the holocaust 'would have gone down in history textbooks as the story of human ascent if Germany had emerged victorious', needs to be treated with scepticism. For how long, one must ask, can a lie be maintained?

The idea of 'desert' as the universal basis of justice is an ancient one. Plato (in *The Republic*) and Aristotle (in the *Nichomachean Ethics*) traced justice to the ideal of a harmonious public life in which all citizens had their place and each received what was appropriate to that place. In the Enlightenment the focus of morality shifted from society to the individual. Sher (1987) finds the transcendental basis for desert in the Kantian principle of autonomy. He argues that the reason why we consider an outcome deserved is that it flows from an autonomous act, that is an act freely chosen by the actor. Because we are free to choose (and if), we therefore deserve the consequences of our actions. In this way 'deserving' is a way of evaluating the consequences of an act for a person against a Kantian ontological criterion – 'because we *are* free', rather than 'because we *ought to be* free'. It is just and fair because you deserve it. You deserve it because you had the freedom to act otherwise. This Sher terms the 'expected-consequence' account of desert. This account certainly matches some of our intuitions about desert. In law (and not only modern western law) punishment is only deserved if the offender had the freedom to act otherwise. *Force majeur* provides exculpation, as does madness, which is believed to limit the person's real freedom to act.

The moral force of 'desert' can be traced back to human freedom. But the question remains, what is deserved? On the one hand we can say that someone deserves to be punished. On the other we can say that someone deserves a jail sentence. On the one hand we can say that a person accused of a crime deserves to be treated with the dignity of a human being. On the other hand we can say that the accused deserves a fair trial. When we use the word in the first sense we are talking about the transcendental quality of the human person; when we use the word in the second we are talking about the customs of a given society whose purpose is to give effect to 'desert' in the first sense.

The exceptions discussed by Sher are instructive. For instance, a person might incur the predictable consequences of an action but suffer either much more or much less than they seem to deserve. A thief calculates that the chances of getting caught burgling a house are small. He does not in fact get caught, and enjoys the spoils. A motorist knows that the risk of serious accident exists, yet drives. She breaks her spine in an accident which is not her fault and ends up a paraplegic. Neither outcome is deserved. Once risk enters the calculation, desert becomes unhitched from freedom of action and there have to be complicated calculations about what part of the actual consequences are deserved. Since almost any operation involving markets also involves chance and risk, a large part of modern experience lies outside the purview of desert. Or, as Walzer (1983) might say, 'desert' is not an appropriate norm for regulating the sphere of commodities, though it is central to the allocation of honours.

Comparing the case for desert against the case for egalitarianism, Young (1992: 339) concludes that while desert cannot be dismissed as a valid basis for justice, its application is limited. 'It is only where agents in circumstances of fair equality of opportunity can take credit for what they do, that any of the various desert bases can ground justifiable claims of desert'. Moreover, he argues that market-based assessments of the value of an individual's contribution depend on a quite different concept from that of desert, namely that of 'entitlement', which is decided expediently by a society and may frequently conflict with that society's own standards of desert (see Chapter 4). These standards are probably formed in the practice of balancing a number of different desert claims in particular cases: for example, the expenditure of effort, the social value of the target of effort, the motives of the agents and so forth. Societies create the institutional mechanisms to make such judgements, including also, of course, judgements about what is *undeserved* such as extreme poverty.

There is one rather general underlying idea which recurs throughout Sher's exploration of the different intuitive bases of desert. This is the belief in the balancing up of accounts, the belief that if something bad happens to a person, then it should be balanced by something good. This belief, modified by the separate belief that virtue and merit deserve reward on their own account, seems to be behind many of our intuitions about desert. It is certainly behind the retributive conception of punishment, and to some extent supports the award of honours for self-sacrificing public service. It is to be found across cultures from the earliest times, very often in the tragedy of its absence – consider, for example, the case of Job in the Old Testament. Most importantly it also supports the principle of compensation, including the idea of a fair wage. Sher expresses the principle as a proposition of 'diachronic fairness', that is fairness over a period of time:

For every good G, every person M, and every period of time P, if M has less (more) of G than he should during P, then M should have correspondingly more (less) of G or some related good than he otherwise should during some later period P.

(Sher, 1987: 94)

In discussions of justice as desert we find a kind of contrapuntal argument between the culturally embedded and institutionally specific conception of desert (*what* people deserve) and a transcendental conception of desert (*that* people deserve something). This argument can be extended to include human-environment relations.

If the foundation of desert is to be found only in freely chosen actions, the concept of *ecological* justice as desert cannot make sense. Neither animals nor rainforests can sensibly be regarded as free or able to choose. Yet, if animals and rainforests have merit or virtue, then desert might apply. If it could be argued successfully that the non-human world has 'intrinsic value', that is value independent of its value to humans, then there would be no problem in justice as 'desert' in a strong sense being applied to human-nature relations. Non-human nature would then deserve to be treated justly because it is an end in itself. If ecological justice as desert were to become a principle, then how that principle is to be observed would vary among cultures and their different institutional mechanisms for finding justice. It is still ultimately human society that decides the standards of merit and virtue embodying the principle of ecological justice as desert. The practical questions cannot be entirely divorced from questions of principle because a principle without any possible practical effect is empty of meaning. *Environmental* justice as desert makes sense so long as the those who suffer bad environments have the capacity to choose, that is so long as they possess autonomy. They might, for example, trade off a bad environment against monetary rewards. But as we have seen in the case of the Ok Tedi mine, this autonomy is in doubt to say the least.

Diachronic fairness seems particularly applicable to ecological justice. We can consider the balancing of accounts with Nature as a matter of desert. In an argument which extends intrinsic value to non-human nature, we could say that the planetary ecosystem does not deserve to be prematurely destroyed. This is not a matter of preserving the planet for future generations of humans but rather of compensating Nature for the destruction caused by past generations. A principle of balance, as Bhaskar (1993: 269) indicates, would place a limit on the emergent totality. If humanity continues to exploit the natural world without giving back what is necessary to maintain the global ecological balance, then humanity deserves the consequence of a degraded environment which may end in human destruction. If this is true, however, we would also have to ensure that the burden of suffering is distributed to people over time in proportion to their contribution to ecological imbalance. Of particular significance here is the environment deserved by future generations. There is a strong basis in desert for stopping activities which load the consequences of overexploitation of the environment on to particular groups and places or on to future generations.

JUSTICE, RIGHTS AND CITIZENSHIP

Justice as respect for human rights has been a major concern of international declarations in the aftermath of the Second World War. Consider the following:

> All human beings are born free and equal in dignity and rights. They are endowed with reason and conscience and should act towards one another in a spirit of brotherhood. (Article 1. Universal Declaration of Human Rights, accepted by the General Assembly of the United Nations in Resolution 217A, 1948)
>
> No-one shall be subjected to torture or to inhuman or degrading treatment or punishment. (Article 3. European Convention on Human Rights, Convention for the Protection of Human Rights and Fundamental Freedoms signed by twenty-one European countries including Turkey). The convention is binding on all member states of the Council of Europe.
>
> Every human being has the right to life, liberty and the security of his person. (American Declaration of the Rights and Duties of Man, approved by the ninth international congress of American states, Bogota, Colombia, 1948). Signatories to the Declaration include all the nations of North and South America.
>
> Human beings are inviolable. Every human being shall be entitled to respect for his life and the integrity of his person. No one may be arbitrarily deprived of his right. (Article 3. Banjul Charter on Human and Peoples' Rights, adopted by the Organization of African Unity, Nairobi, 1981).
>
> It is the duty of every government to insure and protect the basic rights of all persons to life, a decent standard of living, security, dignity, identity, freedom, truth, due process of law, and justice; and of its people to existence, sovereignty, independence, self-determination, and autonomous cultural, social, economic and political development. (Article 1. Declaration of the Basic Duties of Asian Peoples and Governments). The Declaration begins: 'Inspired by the Asian reverence for human life and dignity which recognizes in all persons basic individual and collective rights, rights that it is the duty of other persons and of governments to respect ; . . . '
>
> (de Villiers et al., 1992: 4, 48, 109, 167)

Each of the documents from which the above extracts were taken define in considerable detail the human rights and duties which shall be observed and adhered to by the signatories. They represent an attempt through the United Nations to establish an understanding across cultures of the principle of human rights.

There are, however, considerable differences in the interpretation of the idea of human rights, as Pollis (1992) argues. These interpretations, she points out, are embedded in the different cultural contexts of industrial capitalism, socialism and the Third World. The 1981 African Charter on Human and People's Rights, for example, places fewer restrictions on the exercise of state power than certain other human rights instruments. Certainly the practices of governing elites in the Third World have sometimes not been marked by a high regard for the rights of the individual. The multiplicity of interpretations of rights, she concludes, 'seems to point to a relativistic view of human rights' (*ibid.*: 156). Does it though, when so many authoritarian regimes stemmed not from indigenous cultures but from colonial implantation?

Galtung (1994) is critical of the promulgation of the dominant western view of human rights through the United Nations, claiming that obligations, duties and responsibilities, which are the corollary of rights, become the means for legitimising systems of power embodied in states and corporate capital. The 'fine print' of human rights declarations contains a tissue of omissions and interpretations supportive of existing power structures, including the power to dominate and exploit nature. The human rights perspective, he claims, at least in the limited interpretation given by United Nations declarations, simply does not deal with today's ecological problems arising from human interaction with nature.

The problem, Galtung claims, is structural, arising from myriad acts of commission and omission, none of which can be unambiguously identified as evil. Galtung sees that structural change requires not only change in institutions (courts, legislation, the definition of crime and criminals, actions, actors) but a change in values and understandings which must permeate the world population. It is not that human rights cannot in principle have a universal foundation but that the version of human rights which claims to be universal is in fact rather narrowly based on western cultural norms and power structures.

The emergence of human rights as the basis of justice in western societies historically stemmed from the transcendance of cultural norms. The originators of the idea of natural rights, or human rights, had to side-step their own culture which insisted that all moral values came from God. Both Kant's and Locke's humanist philosophy was anchored in the belief in God, that humans were created by God and that they ought to obey God. But both moved on from there to argue that the law of God is the law of reason. The argument, therefore, did not stop with a holy text but started from that point. The moral law is that which follows from human beings having the (God-given) capacity to reason, that is to observe events, draw conclusions from them and reflect on the consequences of their own and others' action.

Kant's conception of rights stemming from the categorical imperative based upon duty seems to demarcate a rights-based conception of justice which is indeed incompatible with one based on desert or on need. However, the conceptual boundary between rights and need is by no means impermeable. Those who like to refer to Kant (e.g. Hayek, 1960; Nozick, 1974) in justification of the neglect of 'need' by the state have in our view misunderstood him.

It is, of course, true that Kant wanted the duty to others to be founded on a less variable and culturally contingent base than the 'inclination' to acknowledge the needs of others. He writes in *Lectures on Ethics*:

> Let us take a man who is guided only by justice and not by charity. He may close his heart to all appeal; he may be utterly indifferent to the misery and misfortune around him; but so long as he conscientiously does his duty in giving everyone what is his due, so long as he respects the rights of other men as the most sacred trust given to us by the ruler of the world, his conduct is righteous.
>
> (Kant, [*c.* 1780] 1963: 193–4)

Yet, as he makes clear in the following paragraph, it is not provision for needs he is attacking but *voluntary* and *charitable* provision on the basis of emotional inclination:

> Although we may be entirely within our rights, according to the laws of the land and the rules of our social structure, we may nevertheless be participating in general injustice, and in giving to an unfortunate man we do not give him a gratuity but only help to return to him that of which the general injustice of our system has deprived him. For if none of us drew to himself a greater share of the world's wealth than his neighbour, there would be no rich and no poor. Even charity therefore is an act of duty imposed upon us by the rights of others and the debt we owe them.
>
> (*ibid.*)

Kant goes on from there in *Metaphysics of Morals*, Doctrine of Virtue: 'On the Duty of Beneficence' to prescribe the duties of the commonwealth (the state) in respect to meeting the needs of the poor:

> For every man who finds himself in need wishes to be helped by other men. But if he lets his maxim of being unwilling to help others in turn become public, that is makes this a universal permissive law, then everyone would likewise deny him assistance when he himself is in need, or at least would be authorized to deny it. Hence the maxim of

self-interest would conflict with itself if it were made into a universal law, that is, it is contrary to duty. *Consequently the maxim of common interest, of beneficence toward those in need, is a universal duty of men, just because they are to be considered fellow men, that is rational beings with needs, united by nature in one dwelling place so that they can help one another.*

(Kant, [*c.* 1798] 1991, §30: 247, emphasis added)

In §C of 'General Remarks on the Effects with Regard to Rights That Follow from the Nature of the Civil Union' in *The Doctrine of Right*, he argues that the wealthy have an obligation to the 'commonwealth'. This duty can be discharged by, 'imposing a tax on the property or commerce of citizens, or by establishing funds and using the interest from them, not for the needs of the state (for it is rich), but for the needs of the people' (Kant, [*c.* 1798] 1991: 136). It is no exaggeration to say that this establishes the ethical basis of the welfare state.

Kant certainly argued that kindness, love, compassion, and every aspect of empathy was too weak a basis on which to found an ethical system. But this is not, *pace* Okin (1989), to deny affect. The point Kant wanted to make was *not* that such feelings were irrelevant but that they were likely to vary among persons. Some might care, feel empathy and love, others 'couldn't care less'. Caring for the vulnerable in society should not have to depend on the personal inclinations of the healthy, rich and strong. Whether the strong experienced empathy or not, they still had to support the weak. This requirement is derived *not* from any contingent characteristic of persons, be it empathy or rationality or even consciousness, but from the axiom that persons are ends in themselves who require no further justification. From this axiom Kant, in fact, raised 'care' to a principle of distributive justice based on rights.

Thus, the principle of justice according to rights merges into the principle of justice according to needs. The right to be protected from fundamental harms is the same as the right to the satisfaction of basic needs. Indeed, this term 'the right to need-satisfaction' is used by Doyal and Gough (1991) in developing a conception of justice based on needs. Or, as Galtung (1994: 70) argues, rights are the means, and the satisfaction of needs is the end.

Shue (1980) draws on the dual Kantian conceptions of rights and duties to argue that 'basic rights' (for example, to security, subsistence, participation and freedom of movement) have priority over the satisfaction of lesser person-centred rights such as 'preference satisfaction' and 'cultural enrichment'. Shue is writing about the duty of the government of an affluent nation like the United States to work to ensure that people's basic rights, both internally and internationally, are met. Some sacrifice is entailed on moral grounds, but, he says, 'I believe that in fact it is most unlikely that anyone would need to sacrifice anything other than preferences, to which one has no right of satisfaction and which are of no cultural value, in order to honor everyone's basic rights' (*ibid.*: 114). Shue observes:

> I can think of no reason why anyone owes life, fortune, honor, loyalty, or much else to a comprehensive political institution that cannot refrain from or prevent deprivations of security, subsistence, participation, freedom of movement, and the substance of any other basic rights.
>
> (*ibid.*: 113)

Both Shue (1980) and more recently Cummiskey (1996) derive a two-tier consequentialist theory of justice which is consistent with Kant's conception of persons as ends-in-themselves, one in which 'the maintenance of the conditions necessary for rational agency [basic needs] takes precedence over the mere satisfaction of desire' (Cummiskey, 1996: 16, parenthesis added).

Kant ([1798] 1991: 87) wrote that, 'a state is a union of a multitude of human beings under laws of Right', and this multitude and state occupies a certain territory. People and state together, to which the people have consented, is a political community, a 'commonwealth'. Membership of that political community confers rights. Kant viewed these rights as attributes of the citizen which were 'inseparable from his essence (as a citizen)'. The question of what actually constitutes these attributes has been the subject of struggle and debate. Marshall (1950) believed that citizenship involved three forms of rights: civil, political and social. Civil rights of the sort that Kant and Locke discussed came first chronologically, followed by political rights, that is the citizen's rights to play a central role in choosing the legislators. Social rights – the rights to care and compassion – which were present in feudal communities (and, we have suggested, were acknowledged by Kant), in practice sank to vanishing point in the eighteenth and early nineteenth centuries. Only in the twentieth century, with the growth of the welfare state, did social rights attain equality with the other two forms. For Marshall, the principle of citizenship based on the equality of persons opposed the fact of social class based on inequality. The development of citizenship rights had accompanied and inspired the movement towards greater equality which characterised the first three-quarters of the twentieth century: 'the urge forward along the path thus plotted is an urge to a fuller measure of equality, an enrichment of the stuff of which the status [of citizen] is made' (Marshall, 1950: 29; see also Mead, 1986). The European Social Charter of 1961 and the Social Chapter of the Maastricht Treaty are all about social rights.

As we shall see in Chapter 5, the idea of environmental justice as it is being promulgated in the United States contains the postulate of environmental rights. Brechin and Kempton (1994) observe that a number of groups in the environmental movement in the Third World 'are reframing the traditional social justice discourse, such as inequitable development policies, into environmental terms'. The Chipko movement in India, the forest road blockade by the Penan people in Sarawak (Malaysia), the Yanomano resistance to the destruction of their land in the Brazilian rainforest are some well-known examples. Eckersley (1996: 220) asks, 'Would an environmental rights discourse provide perhaps a fourth generation of human rights that might also serve to recontextualise and qualify existing human rights in ways that reflect the late twentieth century political revolution and philosophy of environmentalism?' If so, such a discourse must surely also include the question of the rights of non-human nature as Regan (1983) for one has already foreshadowed. Such a move, however, would require a major reconsideration of the bases of justice.

Recent writers on citizenship have begun to address the issue of how responsible, active and productive citizenship can be encouraged (see Kymlicka and Norman, 1994 and Bookchin, 1990; 1995a, for a more radical view). From the point of view of the environment, it may seem that 'good citizenship' is necessary if far from sufficient to protect environmental values and ecosystems. Bookchin argues for a return to the active citizenship of the small community:

> The classical idea of the rational citizen, engaged in a discursive, face-to-face relationship with other members of his or her community, would acquire economic underpinnings as well as pervade every aspect of public life. . . . It is not too fanciful to suppose that an ecological society would ultimately consist of moderately sized municipalities, each a commune of smaller household communes or private dwellings that would be delicately attuned to the natural ecosystem in which it is located.
>
> (Bookchin, 1990: 194–5)

Different writers have looked to the market, to the education system, and to membership of cultural groups to teach people 'civic virtue', including environmental virtue. But the argument for citizenship confronts the difficulty of inculcating moral virtue without overbearing and oppressive authority. Certainly moral questions, including environmental morality, must be part of political discourse (see Barry, 1996; Christoff, 1996). As Barry (*ibid*.: 128) argues, the politics of sustainablity 'express the attempt to cope with the contingencies of "being in the world", that is being a "citizen-in-society-in-environment"'. In generating appropriate policies which encourage responsibility without being oppressive and unjust, subtle balances must, however, be struck within polities and cannot be legislated in advance. Good citizenship must be regarded as the end result of a political system rather than its foundation.

Turner (1993), therefore, rejects citizenship as a basis for justice. But he argues that the increasing global integration of society both demands and makes possible the development of universal moral principles – that is principles which apply to the whole population of the planet and to all of its various institutions – different in different cultures. Conceptions of universal human rights, rather than ideals of citizenship, are needed now both to protect individuals from oppressive collective institutions and to cope with those situations where the power of the collective to guarantee individual security has broken down. The planetary environment is in need of protection.

Here we are in the presence of a tragic paradox which is at the core of Turner's argument. For the collective institutions human beings set up to enable the pursuit of human ends require a power which invariably leads to the corruption of the ends pursued. We have seen this repeatedly, Turner argues (*ibid.*: 501–3), in the bureaucratisation of church and state and the creation of repressive ideological apparatuses. Even the market, whose ideology appeals to the protection of individual choice, ends by concentrating and entrenching the power of vast corporations to constrain and shape choice for ends dictated by the needs of the corporation and its managers. Following the lead of Nietsche and Weber, Turner concludes that the problem is not one of wrong ideology (e.g. communist or liberal) but is inherent in *all* human collective action. Hence, there is a need to maintain a conceptual sphere of rights which exists outside and beyond the particular culturally relative forms taken by human institutions. Can we find universal features of the human condition which might support a universally applicable conceptual lever for the critique of institutions?

Turner posits three universals: the frailty of the human body, the precariousness of social institutions and the experience of 'moral sympathy'. These universals extend Kant's original perception. The first of these derives from the observation that, at different periods of our life cycle, human beings are totally dependent on others for their life: 'Human beings are frail, because their lives are finite, because they typically exist under conditions of scarcity, disease and danger, and because they are constrained by physical processes of ageing and decay' (Turner, 1993: 501). Even as humanity generates the potential for the abolition of scarcity, it also generates new risks and dangers for particular sections of societies, and across national societies and the world (see Beck, 1992). Social institutions designed to address the frailty of the body are 'precarious' for the reasons already stated. But, again, why should we care about others? Turner answers that:

> human beings will want their (own) rights to be recognised because they see in the plight of others their own (possible) misery. The strong may have a rational evaluation of the benefits of altruistic behaviour, but the collective imperative for other-regarding actions must have a compassionate component in order to have any force.
>
> (Turner, 1993: 506)

Galtung (1994) shares Turner's concern to find a deeper foundation for human rights, one capable of encompassing the human rights bases to be found in other than western cultures, and permissive of different institutional means for their pursuit.

Clearly the theme of 'precariousness', 'frailty' and 'dependence' can be extended to the relationship between human beings and the rest of nature. The constraints under which all of humanity lives are still, as ever, imposed by the natural environment. The success of humanity in conquering natural constraints is a dangerous illusion. Human technical skill and the social institutions to harness that skill have over centuries been devoted to *using* nature. Today the capacity to use nature has developed to such a degree this use threatens to destroy nature. Young and Sachs (1995: 79) point out that 'mining moves more soil and rock – an estimated 28 billion tons per year – than is carried to the seas by the world's rivers'. The capacity of nature to resist entropy is being run down by human overuse (Altvater, 1993). So in confronting the limits imposed by human successes in using nature, we must relearn about our own environmental 'precariousness'.

The place of rights in the discourse of environmental and ecological justice will continue to be debated. Wenz (1988), for example, seems to be in agreement with our interpretation of Kant. Positive human rights, he argues 'require that people provide assistance to one another rather than merely leave one another alone' (*ibid.*: 111). However Wenz does not build his theory on the obligations of society but on those of individuals. Human individuals are morally obliged to protect the positive and negative rights of others. The strength and number of an individual's obligations to others, he argues, are directly related to the 'closeness' of that individual's association with the other. So obligations extend outwards in concentric circles from the individual – with very strong obligations to one's immediate human circle, weaker obligations to other human beings, and yet weaker obligations to animals and non-sentient components of the environment. Such a 'concentric circle' theory is of course exactly the sort of variable conception of justice based on compassion that Kant wanted to avoid. Wenz leaves out of the calculus the fact that human societies devise means of acting collectively. It is the potential for collective human action, public action in the public sphere (both by states and by corporations), that poses the most trenchant questions of environmental justice.

Benton (1993), on the other hand, has mounted a substantial critique of universal human rights with particular reference to the human relationship to nature. His sociological critique points to the inherent social bias resulting from inequality. The rights to property and to communication are of little value to those who do not own property or the means of communication. Moreover, where basic rights conflict, as between the rights of workers to organise collectively and the rights of employers to the beneficial use of their property, and positive law adjudicates, 'the likelihood is that socially, economically and politically powerful groups will be successful in ensuring that their interests predominate in the drawing of the "boundary" between their liberties and those of the less powerful' (*ibid.*: 116).

An individualistic conception of rights, he posits, has not served to protect people from harm arising from three sources: from events where individual responsibility is diffused within an institution such as a state or corporation, from health and other risks arising from an individual's structural class position, and from natural disasters. Finally an individualistic conception of rights, Benton argues, tends to marginalise discourse about a form of society in which the need for protection is reduced. While he does not accept the Marxian vision of a post-scarcity and post-conflict society, he does want to refocus attention from the means of righting current wrongs to acting on the *causes* of those wrongs.

If there is to be a rights basis for environmental and ecological justice, we have to address two questions: Does nature have rights? And do people have a right to their environment? If the answer to the first question is 'yes', then it follows that people have responsibilities to nature. As was argued in Chapter 1, however, the two questions must be considered together because people, 'humanity', is neither a unity in place nor time, and, if 'we' have responsibilities to nature the question immediately arises of how those responsibilities are to be shared among people.

The discussion of justice as rights, like that of justice as desert creates a two-level counterpoint between human rights in principle and the interpretation of rights in different cultures. This counterpoint will continue in any extension of the discussion to human–environment relations. While the question of the distribution of environmental quality can readily become part of the discourse of citizenship, that of the rights of nature cannot be answered by pursuing a Kantian line of argument. As we saw, the argument for human rights depends on our perception that human beings are bodily equal, and to adopt Turner's variation, equally fragile. But the idea of 'equality' makes little sense in describing humanity's relationship with nature. It makes little sense even in describing the relationship between human individuals and individual parts of nature, for example animals. Am I, a human, worth more than a dolphin, a tiger, a slug? The question is vacuous. An individualistic approach does not seem to make much sense because 'the individual' is a concept which seems to have real meaning mainly, though not exclusively, for human beings (but see Regan, 1983: 78–81).

JUSTICE AND NEEDS

As we observed above, the principle of justice according to rights merges into the principle of justice according to needs. So what are needs? Galtung (1994) provides a useful classification of needs into four basic types: (i) material needs which are dependent on individual actors, for example, the need for protection from violence from other individuals; (ii) material needs which are dependent on social structures, particularly the need for 'well-being' or to be protected from physical misery; (iii) non-material needs dependent on individual actors, typically freedom from repression; and (iv) non-material needs dependent on social structures, such as the need for identity. Galtung acknowledges that the divisions are far from watertight but he suggests that the human rights tradition has overlapped more with actor-dependent needs.

The modern idea that the justice of a society depends on that society's capacity to meet human needs begins with Marx. For Marx, needs are discovered by the processes of society, particularly the process of material production. Thus needs are inseparable from the means of satisfying them. The more that a society finds ways of satisfying needs, the more that it produces means for their satisfaction and the more it discovers new needs. If, as Marx says, 'man is rich in needs', this can only mean that people have a rich capacity to conceive of their needs and work to satisfy them. But that capacity is channelled by the system of production which functions by generating and meeting needs. Capitalism is a system in which the discovery of needs reaches new heights (see, for example, the famous passage of *Grundrisse* in the 'Chapter on Capital', Marx, [1857–8] 1973: 409). Capitalism has an almost infinite capacity to develop new *objects* to possess, and thus to develop consciousness of a certain category of need. But this capacity is limited to objects which can be quantified and purchased: 'The need to have is that to which all needs are reduced. . . . It is a need directed towards private property and money in ever increasing quantity' (Heller, 1974: 57). Capitalism is therefore a one-sided mode of production which inhibits the development of a consciousness of those needs which cannot be quantified or marketed.

Eventually, however, Marx believed this needs consciousness would be transcended as a result of the contradictions within capitalism itself. For Marx, the central contradiction was the conflicting development of the forces of production and the relations of production. Heller explains:

> For a certain period capitalism develops the productive forces to an extraordinary degree, through the socialisation of production. Then the socialised productive forces enter into contradiction. This contradiction sharpens, becomes irreconcilable and finally reaches the 'point' at which the centralisation of the means of production breaks the 'shell' of capitalism.
>
> (Heller, 1974: 78)

The capitalism of Marx's day seemed to be following a pattern in which the ever-growing discovery/satisfaction of the needs (quantifiable and marketable) of a relatively small class of property owners could only be pursued via the creation of an organised but impoverished proletariat. The latter would become conscious of its needs in terms of an entirely different and new kind of production system: communism. The development of communism would bring into view the full range of human qualitative needs and the means for their satisfaction.

Though Marx was aware of the possibility that capitalism would find the institutional means to overcome this particular contradiction, he could not have foreseen the extent to which capitalism would absorb a socialist consciousness in order to do so and yet retain its fundamental principles of organisation. The consciousness that societies ought to meet basic human needs, that social institutions must be devised for doing so, that human beings have a 'social right' to have their basic needs met, hence social charters and welfare states, in fact all the adaptive mechanisms which the Right has in recent years tried to wind back, all these display the changed consciousness produced by the social contradiction of capitalism.

The development of fundamental thought about a new kind of society embodying the maxim 'from each according to his ability, to each according to his need' was short-circuited by the real expedients of state socialism and welfare capitalism. But the decline of these expedients has provided a new stimulus for thought, as is evident in the work of scholars like Doyal and Gough (1991) and those who more explicitly address the antinomy within capitalism between the exploitation and degradation of the environment (Altvater, 1993; Beck, 1995; Benton, 1993; Galtung, 1994; Gorz, 1994; Lipietz, 1992; 1996, to name but a few). Though this new thinking cannot be ascribed to proletarian consciousness it is consistent with Marx's general prediction about the emergence of 'radical needs': environmental needs (see Heller, 1974: 86). Galtung's description of the process of emergence of needs is useful here:

> The satisfaction of needs has similarities with political processes in general: there must be some consciousness of the need in the individual; this consciousness must become social and lead to some form of organization through mobilization; there is often some kind of confrontation to have the needs recognized; a real struggle to have the need satisfied may follow; and finally some form of transcendance whereby the need is satisfied individually and its sustained satisfaction more or less guaranteed/institutionalized socially.
>
> (Galtung, 1994: 57)

Even though there is some considerable overlap between rights and needs bases of justice, there is an important practical difference of emphasis. The origin of the human rights base lies principally, as we have seen (*pace* Turner), in perceptions of human autonomy: the capacity to think, reflect and act independently. The origin of the human needs base lies in perceptions of human vulnerability and capacity for suffering. Galtung (1994) argues that human needs are situated *within* the individual whereas rights are located *between* people in that they concern relationships. Galtung further points out a variety of discrepancies between rights and needs. However,

Doyal and Gough seek to bring these two perceptions together under the rubric of needs. The right of autonomy is subsumed within a discourse of 'harm'. They say:

> It is difficult to see how political movements which espouse the improvement of human welfare can fail to endorse the following related beliefs:
>
> 1 Humans can be *seriously harmed* by alterable social circumstances, which can give rise to *profound suffering*.
> 2 Social *justice* exists in inverse proportion to serious harm and suffering.
> 3 When social change designed to minimise serious harm is accomplished in a sustained way then social *progress* can be said to have occurred.
> 4 When the minimisation of serious harm is not achieved then the resulting social circumstances are in conflict with the *objective interests* of those harmed.
>
> (Doyal and Gough, 1991: 2)

These maxims signal a solid commitment to universalism and a challenge to cultural relativism. At the same time criticisms that measures of need are embedded in cultural norms are acknowledged. The underlying and universal bases of need, it is argued, are twofold: survival and autonomy. The former is then revised as 'physical health' since mere survival does not meet the criterion of avoidance of harm and profound suffering. The second is seen to require three secondary criteria: understanding of self and culture, mental health, and opportunities for 'new and significant action', the latter being 'the necessary conditions for participation in any form of life no matter how totalitarian'.

Beyond these basic needs, intermediate needs (adequate food and water, protective housing, a non-hazardous work environment, etc.) are identified whose satisfaction is required in order to allow basic needs to be met. But why should any needs be met? At first Doyal and Gough seem to be appealing to the capacity for empathy with human vulnerability. But they also advance an argument with a strongly Kantian flavour. The capacity for moral behaviour, they argue, is a defining feature of the human being. We have a *categorical* duty to behave well towards others. It is also a feature of the *social* human being. 'The very existence of social life depends upon our recognition of duties towards others' (*ibid.*: 92). The definition of good behaviour towards others may be left to particular cultures, but if a moral community exists at all then everyone has the right to be able to participate in it. 'Therefore, the ascription of a duty – for it to be intelligible as a duty to those who accept it and to those who ascribe it – must carry with it the belief that the bearer of the duty is entitled to the level of need satisfaction for her to act accordingly' (*ibid.*: 94).

65

Doyal and Gough go on to argue that not only is it necessary to ensure that minimal need satisfaction occurs, but that being 'moral persons' in the full sense of the word requires that people have the opportunity to do their best to achieve their full capacity for moral conduct. Optimal performance requires optimal need satisfaction. They thus reverse the argument of the Right that welfare states prevent people from doing their duty, making them dependent and passive. Rather, welfare frequently falls below a level which would enable all people equally to do their best: 'Especially within a competitive economy and culture, it is irrational to exhort the disadvantaged to do their best to help themselves without making provisions for the need satisfaction which they require to do so' (*ibid.*: 101). The argument could be misinterpreted. It is not a claim that possession of material resources (or the lack of them) is correlated with moral (or immoral) conduct. The claim is that people can only be *expected* to participate fully in a community of reciprocal rights and duties if they can rise above the daily struggle to survive. That requires not only health, but also education and the time and intellectual resources to participate as a citizen. In similar vein Peffer (1990) modifies Rawls's (1971) theory of justice and 'difference principle' to argue that not only liberty but *equal worth of liberty* must be guaranteed such that people not only have freedom but the power to make use of it.

That this axis of debate continues to be relevant confirms Marx's perception that ethical questions arise from the real social and material conflicts faced by a society. There can be no doubt that capitalism was transformed in the post-war years and the agents of that transformation were the working class. New ethical standards were inscribed in societies to take account of the injustices pointed out by the working class. But the outcome was a temporary compromise and not a revolution. The compromise suspended the class struggle for a time but the development of new institutional conditions of global capitalism has led to a weakening of the compromise in the industrialised heartland and new struggles on the periphery. The class struggle continues but with another contradiction unfolding alongside: the environmental antinomy in which capitalism destroys the 'nature' it exploits. The 'gale of creative destruction' blown up by capitalism (in Schumpeter's (1943) phrase) has taken on a new and alarming meaning.

The ethic of need has emerged out of the class struggle. The question today is: what ethic will emerge from the environmental struggle? In so far as the class struggle continues, it will have a 'needs' component: environmental needs. The class struggle is also an environmental struggle. Just as in the nineteenth century, the environmental conditions of those at the front line of production is again in question. The ability to destroy the local environment without cost is a benefit which can be offered to producing corporations. In a competitive world, some localities are forced, out of the need to survive, to offer that benefit to corporations. Others, controlled by short-term capitalist interests, seize the chance to offer it. Against this tendency the maxims of Doyal and Gough can be applied:

1 Humans can be *seriously harmed* by alterable environmental circumstances, which can give rise to *profound suffering*.
2 Environmental *justice* exists in inverse proportion to serious harm and suffering.
3 When social change designed to minimise serious harm is accomplished in a sustained way then social *progress* can be said to have occurred.
4 When the minimisation of serious harm is not achieved then the resulting environmental circumstances are in conflict with the *objective interests* of those harmed.

(Doyal and Gough, 1991)

Environmental justice, read as fair distribution of environmental goods and bads, can be based on a principle of environmental need. The need for a safe and healthy environment is a basic need. Optimum satisfaction of environmental need requires much more than this: access to high quality environments – in whatever way 'quality' is defined within the culture. Moreover, following Doyal's and Gough's line of argument, if an environmental morality is to be inculcated in the population, a culture of care for nature, then optimum access to and understanding of the environment is a requirement. But there is a potential contradiction between different understandings of ecological justice relating to the protection of 'wilderness'. As environmental movements grow, spread and encounter opposition in the twenty-first century we can expect many new needs to be identified, the satisfaction of some of which will be entrenched as rights.

Here we come to another question: the needs of nature in ecological justice. In the same way that rights and needs overlap, with the 'right to need satisfaction', so we can say that if nature has rights, then it makes sense to say also that nature has needs. The question hinges on the human–nature relationship. We take up this question in Chapter 6.

SELF, CULTURE AND OTHERS

The bases of justice, desert, rights, needs, express in different ways two things: that there is a universal moral relationship we share with other humans by virtue of their humanness, but that this relationship has to be interpreted through culturally specific institutions which will vary. Pitkin (1972: 187) observes: 'The concept of justice shares with many other concepts in the region of human action and social institutions what I have elsewhere called a tension between purpose and institutionalisation, between substance and form' (she is referring to her earlier work, Pitkin, 1967). The language we use to describe justice refers to different aspects of a real relationship between self and other which we want to achieve. Sometimes the desiderata are contradictory. Yes, we want promises to be kept, but yes we also want to treat people with compassion. People the world over have struggled to capture what is important about the just relationship and different words are used to point out different aspects of this relationship. Social institutions and practical judgements seek to resolve these differences and to weigh up different demands on the relationship (see Light and Katz, 1996).

The relationship between substance and form is not one of clear and concise demarcation. It is more of a continuum. All access to the 'substance' end of the continuum (the transcendental, the universal) is limited by our culture of which social institutions are a part. The different ways in which ancient Greek and modern Enlightenment thinkers sought to describe the universal illustrates the point. But it seems part of the human condition, at least where there is not a rigid and absolute prohibition, to want to gain access to the 'substance' end, and to struggle to describe it in various ways. The continuum between substance and form is expressed by Tully (1995) as two goods which a constitution should mediate: 'The larger purpose of constitutionalism so reconceived, in addition to the recognition and accommodation of cultural diversity, is to mediate the two goods whose alleged irreconcilability is often seen as the source of current constitutional conflict: freedom and belonging' (*ibid.*: 31–2). 'The aspiration to be free from the ways of one's culture and place, and the equally human aspiration to belong to a culture and place, to be at home in the world' (*ibid.*: 32).

Over culturally specific institutions, and thus over the interpretation of the universal, there is a continual struggle. We do not have unmediated access to the universal, indeed we cannot know it directly – as Kant was to conclude in the *Critique of Judgement* (Kant, [1790] 1892). The dialectical struggle must continue. Brennan (1988: 184) remarks, 'I regard any attempt to capture the truth about human nature in any scheme or theory as almost certainly vain'. Kant would disagree, for although he set himself *against* any closed system of universal ethics, he did not regard the *attempt* to find truth as vain. The attempt, however, must be viewed as a continuing and dialectical process, and the belief that one system of thought captures the truth for all time he would certainly have regarded as vain (see *An Answer to the Question: What Is Enlightenment*, Kant, quoted in Reiss, 1970: 57).

The more that discourse about human–other relationships involves intercultural dialogue, the more that the discussion will move up the continuum from form to substance, from being about the culturally specific to being about the qualities shared by humans. The discourse will move towards what 'our' culture aspires to tell 'both of us' about the human condition. Such a move is inherent in the process of discourse itself, for staying down at the culture-bound 'form' end of the continuum renders dialogue, communication and mutual learning impossible. Unfortunately, unless the cultures concerned are tolerant of difference, the result of avoiding dialogue is all too likely a resort to force and oppression to impose one or the other cultural view.

In making moral judgements we are always obliged to refer to a number of ethical frameworks and to consider what is for the best in the particular circumstances which present themselves. Ben Davy (1996) wants the maximum variety of stories and arguments about justice to be heard. Compassion, he argues, is inclusion of all sides in a dispute. Justice cannot be reduced to a single internally consistent system of thought. We should not forget, however, that arguments about justice are not only about justice within the institutional framework we already have but also about the justice *of* these frameworks themselves. Inclusiveness is not the only principle which ought to be taken into account in the formation of such institutions. The justice of institutions is a theme we will take up in the next chapter.

The cases discussed in Chapter 1 should all be matters for public judgement. The Brent Spar case did not come to judgement. Pressure was brought to bear to prevent the environmentally damaging incident occurring. But many questions remain unresolved. Deciding the matter on an *ad hoc* basis by a mixture of market and political pressures simply postpones the time when rational and ethical considerations are brought to bear on the disposal of large-scale oil production plant. So long as disposal has an impact on the environment, the matter is not simply one of technical rationality. As we have discussed, distributional questions arise: who will suffer a worse environment as a result of disposal? Should a production process be permitted which harms the environment? If so, how much harm is acceptable? Can an economy be devised to ensure that the full costs, including environmental costs, of production are contained within the production process? In the short term it may perhaps be in the interests of corporations to externalise costs (e.g. by dumping). But in the long run it is in the interests of corporations to operate in a predictable context of rules which apply equally to all competitors and which cannot be upset by boycotts and politics.

Nuclear testing raises obvious questions of international justice. A 'live' (as opposed to simulated) nuclear test imposes very specific local damage to nature and major risk to human environments. It is not just a matter of nations preparing for war in order to deter other nations from war. Nuclear testing is a technological manifestation of an economic structure. Nuclear devices are the product of a sophisticated and widespread nuclear weapons industry, which in turn is part of a much larger armaments industry. This armaments industry, which produces, at best, nothing but waste (or so common sense suggests), today contributes a substantial portion of the world's gross product. The governments of the world today spend about $800 billion per year on armaments compared with a mere $16 billion on demilitarisation and peace-building activities (Renner, 1995: 166). The ending of nuclear testing in a widely agreed convention does not dispose of the matter of the environmental damage imposed by the armaments industry. At present the institutional means of debating the issue in terms of global justice are largely unavailable.

In the case of the Ok Tedi mining operation, the conflict between local and national needs was focused by the existence of the democratic nation state – there is of course no focus even for that dialectic where the state is not democratic. There was a tendency for the national need for economic development to take precedence over local needs. This dialectic took precedence even over that between the property rights of local land owners and the property rights of mining companies which became aligned with the national interest. Dialectics of more international and global scope had difficulty finding expression because, with one exception, the institutions which might provide the focus of a dialectic of justice are vestigial, fragmented and limited in scope.

In each of the cases discussed the response was reactive. In two of the cases, the Ok Tedi mine and the French nuclear tests, much damage was already done before opposition had any effect. All of the issues were resolved more through pressure group politics than by an attempt to create fair rules of behaviour for states and corporations to follow which would provide a predictable and agreed framework of law based on principles of environmental and ecological justice. The problem today is not the multiplicity of conceptions of justice but the paucity of bases, the extraordinary narrowness of the conception of justice which actually determines outcomes.

CONCLUSION

Thinking about the bases of justice, the words we use to define aspects of our human relationship with others, is a continuing human process. Different institutional systems have evolved in different cultures to interpret the complex and multifaceted idea of justice. If there is something in the idea that human beings have much in common with one another, then there is value in intercultural exchange of ideas. Mutual learning is possible. It is easy to caricature the image of 'Man' which tends to support acquisitive, competitive capitalism: the rational, calculating, self-contained, isolated male. But there is a truth at the core of this image which such caricatures do not touch: that persons, both male and female, desire a certain freedom to shape for themselves their own image of selfhood which goes beyond that which is given to them by their local culture. Were it not so the individualist image would never have taken hold in the first place, nor, having taken root in a capitalist culture, would it ever be possible to break free of it. If this is an idea which stems from western cultural roots, it is one which has found fertile ground in many other cultures. Likewise ideas of the nature of human 'being' have spread from other cultures to take root in the West. These ideas often contain a less Promethean view of the human condition than our western culture tends to emphasise and it is these views which seem particularly important to an extension of the bases of justice to include the non-human natural world more centrally in an ethic of justice.

Finding justice is a dialectical process which cannot be isolated entirely from strategic power struggles. The question is how can those power struggles be contained? How can a space be created in our institutional structures for discursive argument? Finding a political conception of justice is a limited undertaking which seeks to delineate the institutional framework within which the question of justice can be approached. But here, too, we find a diversity of conceptions. In the next chapter we consider some of these different conceptions of political justice. We will see that in this respect too bringing the human–nature relationship to the foreground demands a rethinking and extension of the frameworks of justice.

4

POLITICAL JUSTICE

It seems to me that a fatal mistake of philosophical thought in our day consists in the conception of a fundamental antagonism between the called-for universalism of a post-Kantian ethics and the quasi-Aristotelian ethics of the good life.

(Karl-Otto Apel, 1990: 34)

INTRODUCTION

In the last chapter we saw that the bases of justice have to do with the constitution of persons in the context of their relations with others. We have seen that 'others' need not be restricted to other human beings. But the principal focus has been the human person. We saw that the bases of justice, while denoting important differences, were overlapping and interwoven. The bases enter what we called a dialectic of justice, which results in the creation of particular social institutions. We now turn to the justice of these institutions, political justice.

Societies and their political institutions always already exist. They have not been designed but have evolved, sometimes with decisive turning points like revolutions, but more often by the gradual accumulation of the outcomes of social pressures and struggles. Throughout the last three or four hundred years the constitution of modern society and the division of powers has been contested with something other than brute force; not that brute force has been absent but it too has had to be justified, or 'legitimated'. Persuasive argument has become part of the shaping of society. This argumentative process involves reflection upon the outcomes and purposes of society. The question of justification of society in terms of the interests of all its members has increasingly arisen. Through processes of interpretation and dissemination, systematic justifications of society become part of the accepted wisdom and are applied by decision makers who contribute to the 'shaping' process.

The principal axis of debate today is about how much the institutions of private property and market exchange should be the dominant embodiment of justice and how much and in what way these institutions should be augmented and regulated by a discursive politics in the public sphere. There is also a tension between ethical rules for social interaction, and ethical rules for political interaction including the rules for the process of rule making. This can be clearly seen in the contrast between the idea of 'catallactic' social interaction governed by rules of market exchange and the 'political liberalism' of Rawls dealing with the justice of political–legislative institutions. We will discuss five theoretical systems for the justification of society: the utilitarian, the theory of 'entitlement', the contractarian, the communitarian, and discourse ethics. They are the major markers in ongoing debates about the justification of modern society and they will continue to be important as the debates turn to include the relationship between human society and its natural environment. These markers introduce some important additional concepts about how justice may be achieved. They also deal in different ways with the problem of arriving at a universal morality in circumstances where deserts, rights and needs conflict.

UTILITARIANISM

In broad terms, utilitarianism is: 'the moral theory that judges the goodness of outcomes – and therefore the rightness of actions as they affect outcomes – by the degree to which they secure the greatest benefit of all concerned' (Hardin, 1988: 21). Benefit was defined by the early utilitarians (particularly by Bentham, 1843) as the balance of pain and pleasure, a simple and universal experience which all people share. Utilitarianism was therefore hailed as the philosophy which finally revealed a universal standard for evaluating public policy. Utilitarianism has been properly called a 'public philosophy' (Goodin, 1995). As a theory of political justice, utilitarianism has brought us the idea that all action in the public sphere, variously described as social or political or governmental action, is to be evaluated according to its effects. This seemingly obvious idea is one most people today take for granted. It seems to be simply common sense. The principle has been absorbed into the culture of western democratic society: public action should be judged by its consequences and these consequences evaluated by their contribution to the public good. Without this idea the environment movement could have made little headway, for the environmentalist critique takes much of its persuasive force from the observation that the consequences of social action are tending towards a very unbeneficial outcome for all concerned.

However, utilitarianism also specifies *how* social action is to be evaluated, and in doing so it becomes a narrower and more focused doctrine. Sen and Williams (1982: 3) explain that utilitarianism has three central aspects. The first is the postulate that the correct way to evaluate any given state of affairs is 'people getting what they prefer'. The only reliable measure of value is the balance of pains and pleasures experienced by individuals; and only the individual can be the judge of that balance, which is expressed as a 'preference', or a ranking of preferences. This ranking becomes a measure of the person's 'welfare'. Second, the public choice of any course of action is to be assessed according to its consequences which are described by the welfare of individuals. Finally, the welfare of society or of any group of people is to be assessed by adding up the individual sums of welfare experienced by each person. Thus utilitarianism can be said to be a welfarist, subjectivist, consequentialist and aggregative philosophy.

Utilitarianism, in the simplicity of its basic tenets has provided a powerful justification for the *idea* of critique and practical reform of societies. However, a form of utilitarianism has also become part of the conceptual apparatus justifying the dominance of private property and market exchange. The market exchange of privately owned commodities, it is argued (see below), is the best (perhaps the only) institution capable of efficiently summing individual preferences into a social utility function. Political interference with this institution merely renders its operation less efficient. This view tends to shut out alternative perspectives of justice which might be allowed to enter through a politics of the public sphere. Utilitarianism thus poses four problems in particular for environmental justice which we will call the problem of measurement, the problem of individualism, the problem of monism and the problem of anthropocentrism. Let us take each in turn.

Because the utilitarian philosophy stresses the consequences of action, much depends on how those consequences are measured. Utilitarianism has had an immense impact on governance of the modern economy, both domestic and international. The standards by which economies are evaluated by those who govern them (national governments, the World Bank, the IMF) depend on one type of measure in particular: gross national product or gross domestic product (GNP, GDP, the difference is significant). Marilyn Waring (1988) points out that the modern concept of GNP as it has been applied in the post-war world can be sourced to a paper by British economists Maynard Keynes and Richard Stone (*c.* 1939) entitled 'The National Income and Expenditure of the United Kingdom, and How to Pay For the War'. This paper was followed by another by Milton Gilbert (1941): 'Measuring National Income as Affected by the War' (these papers are cited by Waring, 1988: 55).

The latter paper is said to be 'the first clear, published statement of Gross National Product' (Duncan and Shelton, 1978). GNP was never intended as a measure of 'welfare' but simply of business activity with a view to determining the national income of a country at war. The wellbeing of a population cannot be and was never intended to be measured by GNP. Moreover GNP was later revised as the primary measure of growth (read 'economic success') and replaced by GDP. Whereas GNP is a measure of production that generates money income for a country's residents, GDP (gross domestic product) measures production that generates income in a nation's economy 'whether the resources are owned by that country's residents or not' (Waring, 1988: 71). Thus income which accrues to non-residents (in the form of profits, interest, etc.) from a given national economy is part of its GDP but not of its GNP. Mining ventures may add to the GDP of a country like Papua New Guinea but may add relatively little to the (unevenly distributed) financial benefit of residents.

GDP includes a large range of activities that we might well think *detract* from the wellbeing of people: the armaments industry, gambling, prostitution, the use of drugs such as tobacco, and the production of hazardous chemicals. The production of new items is included but the *re-use* of products such as automobiles, which might be a more sustainable use of resources, is a minus for GDP. GDP *excludes* an even greater range of activities that *support* wellbeing, including 'domestic' work traditionally (though today not exclusively) done by women. Most importantly from the point of view of environmental and ecological justice, the measurement of GDP excludes from the national accounts the value of the environment. Included in GDP on exactly the same terms are activities that destroy environmental value (e.g. pollution) and those that restore it. As Jacobs (1991: 224) points out, GDP growth is not even an adequate measure of the long-run 'health' of the national economy because, as is now well understood, economic health is dependent on a sustainable economy and GDP growth does not measure resource depletion. It is as though we were to measure the health of a lung cancer patient by the treatments he receives – without mentioning that he smokes forty cigarettes a day.

Jacobs (*ibid*.: 226) also argues that GNP/GDP plays two quite separate roles in public life: as a tool for understanding how the economy works – 'an objective measure without moral content' – and as a tool for measuring the success of an economy. Unfortunately, however, these functions cannot be so neatly separated because measures of how the economy works imply an understanding of what 'the economy' is. If this understanding is deficient or biased to favour particular interests (e.g. business, productivist and male interests), then that bias will be transferred to measures of the economy's success.

Waring's critique does not invalidate the consequentialist principle of utilitarianism because it may be that a better method of measurement can be found. But it does throw doubt on the utilitarian claim to have discovered a universal foundation for political justice. While pain and pleasure may be experienced by all humans equally, this observation can have no impact on society unless the pains and pleasures can be measured. Measurement is culturally determined. Indeed the claim that statisticians can rise impartially above the political fray and provide a value-free account of costs and benefits is, to use Bentham's language, 'nonsense on stilts' (see Self, 1970).

Beck (1992) has shown how the argument over the measurements of effects, and now over the *risks* of effects occurring is today heavily politicised. Moreover, the reduction of environmental values to what is easily measurable through the medium of money, and the trading of those money values, is highly problematic as we discuss in the next chapter.

The second and third problems are closely related. Utilitarianism conceives of persons rather narrowly and acknowledges only one best way of aggregating preferences. The well-known claim of utilitarianism is that individuals are the best judge of their own best interests (see Mill, 1971). In some libertarian interpretations this postulate becomes the maxim that whatever wants individuals express they should have. Goodin (1995: 119–31) has discussed this issue at length. As he points out (*ibid.*: 130), persons are often conflicted as to their own interest – use of drugs, gambling, overeating, driving with alcohol in the blood. They may know that stopping these habits is in their best interests. But they may still continue. Elster (1982) argues cogently that what people want is quite strongly influenced by what is made available to them. His point is that we have to take account of how wants are socially generated. The direction of these critiques is that the picture of the isolated individual is not a true reflection of reality. Persons define themselves and what they want continually and discursively in interaction with each other at many levels – in families, in groups, and at the level of public politics. Even as we continue to desire the goods provided by today's capitalist system, we may yet realise that the consumption of these goods is not in our best interest and we may simultaneously seek to reform the system. But this process of reform can only be progressed by joining together with others in political action.

On account of its consequentialist principles, utilitarianism tends to regard justice as lying essentially in finding a single 'best' outcome. Arguing that persons are individually the only competent judges of their best interests, utilitarians nevertheless seek institutional rules for aggregating those individually defined interests into a single optimum. In an important critique Rawls (1982) shows that utilitarians assume that all conceptions of the good held by persons are commensurable. This assumption is made plausible by reducing conceptions of the good to subjective tastes or 'preferences' which can all be aggregated and subsumed under a 'shared highest-order preference function' (or social utility function). But in doing so the idea of 'the person' becomes impoverished. Rawls comments:

> This loss of individuality suggests that the notion of a shared highest-order preference defines persons as what we may call 'bare persons'. Such persons are ready to consider any new convictions and aims, and even to abandon attachments and loyalties, when doing so promises a life with greater overall satisfaction, or wellbeing, as specified by a public ranking. The notion of a bare person implicit in the notion of

shared higher-order preference represents the dissolution of the person as leading a life expressive of character and of devotion to specific final ends and adopted (or affirmed) values which define the distinctive points of view associated with different (and incommensurable) conceptions of the good.

(ibid.: 181)

Utilitarianism can thus become a monistic and indeed somewhat authoritarian doctrine which simply refuses to countenance more than superficial difference among persons. Concern for the environment becomes simply another taste which can be bought off and traded against other tastes under the automatistic rule of particular institutions such as those of the market which are supposed to be the best mechanism for aggregating tastes into a higher-order preference function. As we shall see (Chapter 6), bringing the environment into moral considerations, pursuing environmental and ecological justice, requires an expanded conception of the horizons of the self. Rawls (1982: 183) counterposes an idea of society in which the members are 'conceived in the first instance as moral persons who can cooperate together for mutual advantage, and not simply as rational individuals who have aims and desires they seek to satisfy'. The idea of a moral person is one whose personhood is defined in some way with respect to others and to the natural world around. People are more than the aggregate of their wants. While the market is a useful instrument for serving human wants, it is not an instrument for expressing, debating and deciding upon the higher ends of humankind which also contribute to each person's 'personhood'.

Finally, utilitarianism is an anthropocentric philosophy. The use to which it has been put is the critique, reform and justification of human institutions involving human agents. Now it is true that the founder of utilitarianism, Jeremy Bentham, argued that animals appear to experience pain as much as humans and therefore the utilitarian calculus includes animal welfare. Singer (1979) has made much of this point. Animals he argues cannot be considered to be moral 'agents'; what animals *do* cannot be considered in moral terms, but what they *suffer* is a matter of morality. Animals are moral 'patients'. Singer (1979: 13) claims that, 'The utilitarian position is a minimal one, a first base which we reach by universalizing self-interested decision making'. He then makes the case on utilitarian grounds for including animals among those persons on whose behalf ethical decisions are made by others: children and those humans considered unable to make decisions in their own best interest. As is well known, Bentham applied his critique to the kinds of human institutions which have a duty to care for those considered incompetent to make decisions on their own behalf: prisons, for example, mental institutions or schools.

Treating animals as one would treat the occupants of prisons, mental institutions and schools is certainly a major advance (from the animals' viewpoint) from treating them as inanimate objects. But the approach begs the important moral question of whether animals should be treated according to the standards we apply to humans or just left alone to experience the treatment nature hands out to them. The socially constructed standards of measurement of pain and pleasure again enter the calculus. How much is pain/pleasure a brute fact, and how much a socially constructed condition? There seems to be an important distinction here between domesticated and wild animals. The natural condition of domesticated animals, evolved over millennia, is to be cared for by humans; the same cannot be said for 'wild' animals. Moreover, while a utilitarian approach might apply to the higher animals which can be said to suffer (really only a small part of the biosphere), it does not apply to the parts of nature which cannot meaningfully be said to experience suffering. These are matters we take up again in Chapter 6.

In spite of these problems, the consequentialist principle of utilitarianism 'as a public philosophy' will continue to be of great importance as the world struggles with environmental and ecological problems. Attempts to broaden utilitarianism and restore its critical thrust continue (see Sheng, 1991; Häyry, 1994). Needs and 'the quality of a social distribution' in terms of fairness and equality have been brought into the calculation of social utility. As Goodin explains:

> Utilitarians are outcome oriented. In sharp contrast to Ten Commandments-style deontological approaches, which specify certain actions to be done as a matter of duty, utilitarian theories assign people responsibility for producing certain results, leaving the individuals concerned broad discretion in how to produce those results. . . . The distinctively utilitarian approach, thus conceived, to international protection of the ozone layer is to assign states responsibilities for producing certain effects, leaving them broad discretion in how they accomplish it.
>
> (Goodin, 1995: 26)

The measurement of outcomes will continue to be crucially important in directing public policy for the environment. The fact that standards of measurement are today themselves subject to political debate and critique merely reinforces the importance of such measurements. Perhaps also the contrast between consequentialist and deontological approaches is not as sharp as Goodin suggests. Why should we measure outcomes? Because people have rights and needs which must be met. Why should we leave people a wide area of discretion to decide their own ends? Because people are to be treated as ends in themselves. Why should we regard animals as persons? Because they share with humans the property of being 'ends in themselves'. Extending such insights to the world beyond humanity, however, raises questions we will consider later.

ENTITLEMENT

A different justification for existing institutions of private property and market exchange is provided by the theory of 'entitlement'. In this theory, in contrast to utilitarianism, consequences, outcomes, are not to count in the calculation of justice. The theory reaches its zenith in the work of Nozick (1974). Nozick's entitlement theory of *Anarchy, State and Utopia* is not the product of his mature thought. In *The Examined Life* (1989) he substantially retreats from his unqualified defence of unrestrained capitalism and the minimal state. However, the former remains an influential doctrine in today's political environment. Nozick's work is an interpretation of the works of Hayek (1944; 1960; 1976; 1979), who in turn bases his conception of justice on an interpretation of the work of Kant and Locke.

Entitlement theory, as Nozick explains is an historical theory of property. All we need be concerned about is the justice of original acquisition of property and the justice of its transfer. If we assume that property was justly acquired, and if the exchange of property is based on the principle of the contract, with mutual promises between two parties, then we have no further need to be concerned with the justice of outcomes. Nozick writes:

> In a free society, diverse persons control different resources, and new holdings arise out of the voluntary exchanges and actions of persons. There is no more a distributing or distribution of shares than there is a distributing of mates in a society in which persons choose whom they will marry. The total result is the product of many individual decisions which the different individuals are entitled to make.
>
> (Nozick, 1974: 149–50)

Liberty, Nozick says, upsets social patterns such as 'equality'. Those who seek to redistribute resources by, say, taxation undermine each person's liberty to spend their own resources on themselves or others. Inequality, and thus riches and poverty, arises from people being free to choose to whom they transfer the property they have legitimately acquired either by production or exchange. People are entitled only to what they produce or acquire through a contract of exchange, and they are entitled to dispose of their property as they wish. This is the principle of entitlement which underwrites both justice in acquisition and justice in transfer.

Entitlement theorists are of course concerned about the justice of the transactions through which property is exchanged. Four assumptions are entailed. First, it is assumed that the law of contract, and its policing, will be sufficient to uphold the justice of transfer. Second, it must be assumed the ownership of property confers no power which can have an unjust effect on the outcome of bargaining over a wide range of political matters not directly involving an exchange of property. Third, luck, including the luck of being born with particular characteristics, is discounted. Moreover the cumulative effects of luck – either good or bad fortune – are not the concern of justice. Finally, it must be assumed that the original acquisition of private property was just.

Hayek, reacting against the state-organised collectivism of fascism and Stalinism, and what he saw as the corruption of liberal society by its influence, wanted to reinstate the individualism of Kant and Locke:

> The respect for the individual man *qua* man, that is the recognition of his own views and tastes as supreme in his own sphere, however narrowly that may be circumscribed, and the belief that it is desirable that men should develop their own gifts and talents.
>
> (Hayek, 1944: 20–1)

The institutions of the market: property, the price mechanism, money, are postulated as the great social discovery which reconciles individual freedom with the need for social co-ordination (Mises, [1949] 1963 ; Hayek, 1976). These institutions, it is argued have rendered the discourse of social justice obsolete.

What Hayek makes clear is that there is no *intentional* or *foreseeable* distribution on the part of individuals. When we buy or sell something we do not consider the effect of that action in terms of how rich or poor it makes certain other people in society. Moreover, because the relevant knowledge is not available, we cannot *know* what effect our action has on the distribution of resources within society at large (e.g. what contribution to employment, profits, riches, poverty, etc. in what sector of society). The validity of Hayek's argument against social justice depends on the claim that politics has no place in determining distributions in a modern market society. Hayek does not claim that the state has no place, but that the role of the state should be restricted to upholding the law necessary for the price mechanism to work with maximum efficiency, in effect the state upholds the rules of a 'game'. The game of 'catallaxy', Hayek states:

> proceeds, like all games, according to rules guiding the actions of individual participants whose aims, skills, and knowledge are different, with the consequence that the outcome will be unpredictable and that there will regularly be winners and losers. And while, as in a game, we are right in insisting that it be fair and that nobody cheat, it would be nonsensical to demand that the results for the different players be just. They will of necessity be determined partly by skill and partly by luck.
>
> (Hayek, 1976: 71)

These ideas have been applied to environmental questions. Ackroyd *et al.*, commenting on the New Zealand experience, write:

> Many environmentalists have joined economists in recognising the inefficiencies and costs associated with collective decision-making by

government agencies. . . . Markets, based on carefully defined, secure
and tradeable ownership rights, have a greater potential for improving
our quality of life than centralised resource management.

(Ackroyd *et al.*, 1991: 57)

Capitalist markets must be allowed to address the task of cleaning up and
'greening' the planet (Salmon, 1991). The future of the world economy will
see a growing role both for environmentally friendly production techniques,
and for the rectification of currently degraded environments (Elkington,
1987). Biodiversity can best be protected by allocating property rights to
endangered species (Chisholm and Moran, 1993). The 'green economy' will
be a capitalist economy. And just as the economy theoretically reaches a level
of equilibrium in which social needs are met, so the green economy will
theoretically reach a level of 'sustainable development' in which the capacity
of the planet to provide raw materials and absorb wastes is not overstretched.

'Sustainable development' is reconstrued as 'sustainable management' by
producer corporations. The role of government is not to concern itself with
outcomes but to protect property rights. The environmental problem is
taken to lie with political decisions supplanting market decisions and with
government intervention leading to:

> arbitrary actions based on political rather than scientific reasons,
> unduly high uncertainty surrounding development approval processes
> and reduced investment in resource based industries, inflexibility in
> negotiating mutually beneficial trade-offs between conservation inter-
> ests and commercial development, and the costs of conservation being
> borne by specific industry sectors, rather than the community that
> benefits from conservation.

(Chisholm and Moran, 1993: 160)

Entitlement theory, since it is not a consequentialist theory, reduces the
utilitarian problem of measurement from a fundamental to a contingent one.
But this very aspect makes entitlement theory, at least in its Nozickian
form, inapplicable to the environmental problematic of the planetary future.
If it is accepted that there is a problem of environmental degradation and
resource depletion, then all that entitlement theory can say is, 'This is a bad
problem but it has nothing to do with justice'. There is a tendency for those
who still believe in entitlement theory to argue that we have to wait for the
problem to be confirmed by science before acting. The necessarily collective
action on the problem, entitlement theorists would argue, defeats the principles of justice.

In other respects entitlement theory retains the problematic features of
utilitarianism: of restrictive individualism, monism and anthropocentrism.
Since animals and other aspects of non-human nature cannot own property –
though they can become property – the theory is even more strongly

anthropocentric than utilitarianism. There are also a number of additional difficulties which considerations of environmental and ecological justice pose for the theory. First, it cannot be assumed that the original acquisition of property was just. Second, while to play the catallactic game means to obey the rules, there is also another game of 'making the rules'. Third, property confers political power which is particularly useful when it comes to playing the game of 'making the rules'. Fourth, the concept 'environment' cannot be reduced to the concept 'property'.

The problem of original acquisition was considered by Cohen (1986) in a well-known refutation of Nozick's theory. In Cohen's view Nozick's arguments simply reduce to a justification of the right of the stronger. Cohen points out that Nozick ignores the possibility that property may be originally owned in common rather than not owned. Common ownership implies a quite different and more discursive politics of distribution than individualised ownership. Cohen is talking mainly about land. So also was Kant when he considered the question of original acquisition. Kant considered that a colonising 'nation' may justly occupy *vacant* lands. But he was dismissive of specious arguments in favour of forcible colonisation of already occupied lands. He recognised that indigenous people who are shepherds or hunters depend for their sustenance 'on great open regions' and may occupy the land only thinly (see Figure 4.1). He had in mind the colonised territories of Africa, America and Australia (see Kant, [*c.* 1798] 1991: 159, *Metaphysics of Morals* III §62). Kant argues that these colonial invasions cannot be legitimated in the name of justice.

Kant's theory of property is also quite different from that of Nozick. As Kersting explains:

> The right of reason grounded in freedom demands private property. . . . If the original right of freedom finds its external guarantee in property, then every human must have a right to property grounded solely in the right to freedom, which must be ascribed to him merely on the basis of his humanity. Obviously this conception of the right to property calls for a positive politics of distribution by the state.
>
> (Kersting, 1992: 348)

In fact, Kersting points out, for Kant the state (we would prefer to say, 'the public sphere') is called into being for the purpose of putting into effect a transition from a condition where all property is held in common to one where private property is created. But the justice of this transition lies in everyone affected being in agreement on the distribution at every stage. The right to property cannot, therefore, be reduced to the juridical expression of this right in empirical acts embodying 'entitlement'. Thus Nozick's entitlement theory of justice is not consistent with Kant's theory of property (*ibid.*: 351).

Figure 4.1 An aboriginal family at the time of the occupation of
'New Holland' by the British

Source: 'Smoking out the opossum', illustration from *Field Sports of New South Wales,* by J. H Clark, 1813,
London: Edward Orme

There is a wide range of environmental goods which cannot be reduced to
private property (see Dryzek, 1987). In this respect, Kant's strong anthropocentric outlook can be challenged. Kant meant 'property' to be
understood as the resources which allow the moral being to flourish to its
fullest extent in accordance with its nature. If the absolute limits of the
moral are extended to include non-human nature, then difficult questions

arise about the 'property' of non-human nature. But the mechanism of distribution of 'property' must be decided in the public sphere (in Kant's terms, by 'the state'). The human being becomes something more like a partner sharing nature rather than, as Kant thought, nature's overlord.

Hayek's theory of justice is in some respects closer to Kant's and Bentham's than to Nozick's. Hayek does not oppose regulation of the economy, but insists on the 'rule of law': 'all coercive action of government must be unambiguously determined by a permanent legal framework which enables the individual to plan with a degree of confidence and which reduces human uncertainty as much as possible' (Hayek, 1960: 222). Hayek also discusses what he calls 'neighbourhood effects' in which the use of land by an owner in certain ways has an impact on the enjoyment by neighbouring owners of their own property. He thus implicitly acknowledges the value of an *Umwelt*, an environment, which is quite different from property in land. However, he did not live to see that the 'neighbourhood' for certain effects is now the whole planet, nor that 'environment' is much more than whatever is left over after dividing up nature for human use.

CONTRACTARIANISM

We turn now from systems of political ethics whose main focus is the ethical rules which should govern *social interaction* to the rules of *political rule making*. Let us begin by considering Nozick's mature reflection:

> The capitalist ideal of free and voluntary exchange, producers competing to serve consumer needs in the market, individuals following their own bent without outside coercive interference, nations relating as co-operating parties in trade, each individual receiving what others who have earned it choose to bestow for service, no sacrifice imposed on some by others, has been coupled with and provided a cover for other things: international predation, companies bribing governments abroad or at home for special privileges which enable them to avoid competition and exploit their specially granted position, the propping up of autocratic regimes – ones often based upon torture that countenance this delimited private market, wars for the gaining of resources or market territories, the domination of workers by supervisors or employers, companies keeping secret some injurious effects of their products or manufacturing processes etc. This is the underside of the capitalist ideal as it actually operates.
>
> (Nozick, 1989: 280)

Some might think this an understatement. No mention is made of the environmental damage caused by predatory and invasive capitalism of the kind illustrated in Chapter 1. But it is a courageous admission by a philosopher whose earlier ideological position has become one of the central props of capitalist ethics. Maynard Keynes (1931) in an essay of 1926 entitled 'The End of Laissez-Faire' made essentially the same point. But it seems we have to relearn the lesson at regular intervals.

As discussed above, the test of justice for Kant was whether people would agree to the distributions adjudicated by the state. The contract was not an historical event but a thought experiment functioning as a standard to be constantly applied by rulers to their decisions. The 'original contract' according to Kant occurs the moment a people constitutes itself into a state. It thereafter becomes the standard for all governance. Kersting (1992: 355) tells us that: 'The original contract is the model of a procedure of advice, decision, and consensus that guarantees the justice of its results because these are supported by universal acceptance'. This principle is compelling because the keeping of a promise is part of the practical and linguistic idea of what a 'promise' is. A person would be engaging in a deception (making a 'false promise') if the person did not expect to keep it. If neither party had any expectation that the promise would be kept, they would be engaging in a nonsensical interaction.

Rawls followed Kant in his thought experiment, asking what sort of rule people would impartially agree to when constituting themselves into a political society. His answer is well known:

> Each person is to have an equal right to the most extensive total system of equal basic liberties compatible with a similar system of liberty for all. Social and economic opportunities are to be arranged so that they are both: a) to the greatest benefit of the least advantaged [the 'difference' principle], consistent with the just savings principle, and b) attached to offices and positions open to all under conditions of fair equality of opportunity.
>
> (Rawls, 1971: 302–3)

These precepts are meant to define the foundations of a just society. They are not designed to provide pragmatic, day to day standards for political action.

Utilitarianism, he argued, wrongly placed 'the good' before 'the right'. In whatever way 'the good' may be defined by a particular society (in terms of satisfaction, pleasure, happiness or civic virtue), 'the right', Rawls thought, the rules under which the good is pursued, must take precedence. Rawls (1982:179) specifically includes as a conception of the good, 'attitudes of contemplation toward nature, together with the virtues of gentleness and the beneficent stewardship of natural things'. He also reacts against the metaphysical deontology of Kant, substituting the norm of impartiality for the categorical imperative. Since the 'judges' of the contract are the parties themselves, impartiality must somehow be inscribed in the make up of the parties. This is the intention of the device of the 'veil of ignorance' in which the parties have no knowledge of the effect the institutions of the contract will have on their subsequent welfare (see Barry, 1989: 183). One of the empirical assumptions Rawls makes is that, together with liberty, people would want some guarantee of security in the economic circumstances of their lives. This is the assumption which leads to the 'difference principle'.

Rawls takes the view that individuals cannot be said to 'deserve' their place in society or the 'assets' with which they come into the world: abilities, traits and talents. This is entailed by the presumption of intrinsic equality. Further, if people do not deserve their 'natural assets' then no more do they deserve the advantages that those assets make possible: wealth, status and so forth. So Rawls is opposed to the concept of 'desert' as a principle of justice. Taylor (1986: 58) agrees with him that 'talents' are to a large degree socially created (through education, parentage and the like). Rawls's 'general conception' is that, 'all social primary goods – liberty and opportunity, income and wealth, and the bases of self-respect are to be distributed equally unless an unequal distribution of any or all of these goods is to the advantage of the least favored' (Rawls, 1971: 303).

In the *Theory of Justice* (1971), Rawls is postulating what amounts to a single fixed rule for a 'well ordered society'. The same would be true, however, of the 'well ordered societies' postulated by the philosophies of utilitarianism or entitlement theory. In *Political Liberalism* (1993a), he has moved towards a consideration of the conditions for consensus on a political society in which a number of such rules are legitimately advocated and debated by citizens with competing interests. It is evident to Rawls that no well-ordered society can exist without some degree of consensus on fundamental principles. However, there must be room to debate a variety of what Rawls terms 'reasonable comprehensive doctrines'. It is in this respect that he differentiates his position from Kant's (Rawls, 1993a: 99–101). Kant's transcendental idealism assumes that a single rule must govern both political relations and all of human relations. Different political systems in practice interpret this one transcendental truth in different ways (based on the ideal of autonomous persons as unqualified ends-in-themselves). Rawls turns this idea around. For Rawls there is no single transcendental truth. There are a number of 'truths', but there is a single *political* rule specifying the conditions under which people holding different 'truths' can co-operate for reciprocal advantage. 'Justice as fairness' views society as a fair system of co-operation. Autonomy is to be viewed as a political and not an ethical value: 'It is realized in public life by affirming the political principles of justice and enjoying the protection of the basic rights and liberties; it is also realized by participating in society's public affairs and sharing in its collective self-determination over time' (*ibid.*: 77).

It is natural to believe, Rawls argues, that 'social unity and concord' requires agreement on a comprehensive moral doctrine (*ibid.*: xxv). But the philosophy of pluralism has its roots in the initially contentious communicative practice of toleration. 'Political liberalism assumes that, for political purposes, a plurality of reasonable yet incompatible comprehensive doctrines is the normal result of the exercise of human reason within the framework of the free institutions of a constitutional democratic regime' (*ibid*.: xvi). Rawls then asks: 'How is it possible that there may exist over time a stable and just society of free and equal citizens profoundly divided by reasonable religious, philosophical and moral doctrines?' (*ibid*.: xxv). In order to answer this question we must construct the principles of political justice which make such a condition possible. Rawls is in effect seeking to specify the requisite balance between, on the one hand, *justice*, and, on the other, *tolerance*. In doing so it becomes necessary to make a distinction between justice as realisation of the highest good and a *political* conception of justice. In doing so he moves the dialectic of justice to a new level.

A political conception of justice which reconciles justice and tolerance has three characteristics. First, its particular subject is the political, social and economic institutions which form the 'basic structure' of a modern constitutional democracy. A political conception, therefore, does not claim to refer to wider matters of human relationships or qualities. Second, a political conception of justice is *presented as* 'a free standing view', that is, without necessary reference to any particular comprehensive moral doctrine. It will be embedded within one or more such doctrines (it may be 'part of or derivable within' such doctrines) (*ibid*.: 12). For instance, toleration has at times been part of both Christianity and Islam without changing the core beliefs of either religion. Third, the content of a political conception of justice is expressed in terms of certain fundamental ideas seen as implicit in the public political culture of a society (*ibid*.: 13). The political culture of a democratic society is 'a tradition of democratic thought' familiar to and intelligible to 'the educated common sense of citizens of that culture'.

Central to a political conception of justice is the concept of reciprocity: namely the terms regulating the behaviour of citizens which each citizen can accept, provided that every other citizen accepts them. Reasonable people, Rawls thinks:

> desire for its own sake a social world in which they, as free and equal, can cooperate with others on terms all can accept. They insist that reciprocity should hold within that world so that each benefits along with others.
>
> (*ibid*.: 50)

Rawls insists that rational agents are not narrowly self-interested:

> Every interest is an interest of a self (agent), but not every interest is in benefits to the self that has it. Indeed, rational agents may have all kinds of affections for persons and attachments to communities and places, including love of country and of nature; and they may select and order their ends in various ways.
>
> (*ibid*.: 51)

In social and political life our actions are based not so much on facts as on judgements. We cannot avoid making judgements (for reasons set out in detail in Rawls, 1993a: 56–7). In so far as we make judgements, and feel entitled to do so, we recognise that others are entitled to do likewise. Since we all suffer from the same lack of perfect evidence for our judgements, and since we all differ in the reasonable ways in which we arrive at our judgements, we all carry the same 'burdens of judgement' – the burden of being unsure. Reasonable people thus recognise that others are bound to differ from them in good faith. This recognition leads to tolerance and acceptance of 'reasonable pluralism' under which an overlapping consensus can be achieved among reasonable comprehensive doctrines.

As in *Theory of Justice* (1971), Rawls argues that 'the right' (the justice of how things are to be decided) takes precedence over 'the good' (what is decided). Nevertheless, there are certain substantive conditions necessary for people to have the capacity to make decisions co-operatively within a framework of 'the right'. These conditions give recognition to basic needs (cf. the arguments of Shue, 1980; Goodin, 1988; and Doyal and Gough, 1991). Rawls terms these conditions 'background justice' (Rawls, 1993a: 265–68). In an implicit response to Nozick he gives four reasons why it is the role of institutions of the 'basic structure' to secure background justice. First, the conditions of interaction among individuals are socially given. We have to ensure that the rules of the market are fair: 'excess market power must be prevented'. Second, fair 'background conditions may exist at one time but be gradually undermined even though no-one acts unfairly when their conduct is judged by the rules that apply to transactions within appropriately circumscribed local circumstances'. Third, because rules for social interaction have to be simple and comprehensible, there are no feasible and practical rules that can prevent the erosion of background justice. Finally, because of these effects there must be a 'division of labour' between two kinds of social rules: those governing the maintenance of background justice (e.g. progressive taxation necessary to ensure that everyone has a fair chance to compete in markets) and those governing the interactions themselves (e.g. rules regarding fair competition, fraud, etc.). Thus, Rawls argues:

> What the theory of justice must regulate is the inequalities in life prospects between citizens that arise from social starting positions, natural advantages, and historical contingencies. Even if these inequalities are not in some cases very great, their effect may be great enough so that over time they have significant cumulative consequences.
>
> (*ibid.*: 271)

Rawls's principles of justice are an extension of Kant not a rejection. It is the idea of the right of autonomy and the idea of the human person as an end requiring no further justification which supports the idea of political tolerance and the imperative of co-operation. Rawls's account of political justice takes a major step away from monism and narrow individualism. What becomes central is the political institutions for co-operation, and the conditions under which such co-operation can be achieved. The question is not how individuals can be co-ordinated, but how the varying values of the 'lifeworld' (to use the term of Habermas) can be allowed to penetrate the systems for co-ordination currently in place.

The limitation Rawls places on the scope of his theory of justice raises some important problems. Feminists have pointed out that Rawls excludes the family from the scope of justice (Baier, 1992; Okin, 1992; Kearns, 1992). The family is a site of unjust, gendered power relations. The reason Rawls wants to exclude the family from considerations of his theory of justice is that the latter is meant for a particular purpose: the rules of 'the basic structure', an idea which embodies the division of society into private (family) and public (political) sectors. The basic structure belongs to the latter. Kearns (*ibid.*: 484) asks, 'No-one wants to abolish private life, but can a totally just public structure be erected on an unjust base?' If conceptions of justice other than those specified by Rawls are to regulate *private* life, why should these principles be excluded from *public* life? In fact, as Okin (1992: 554) points out, Rawls himself relies upon human feelings of empathy, benevolence and concern for others 'in order to have the parties [to the contract] come up with the principles they choose, especially the difference principle'. As Baier (1992: 56) concludes, and we agree, there is a 'need for more than justice'. There is a need in public as well as in private life for virtues as well as rights: 'The best moral theory has to be a cooperative product of women and men, has to harmonize justice and care' (*ibid.*).

Rawls has little to say about the environment. From the point of view of environmental justice, there appear to be grounds for including (human) environmental needs and quality as part of background justice. To the extent that humans are dependent upon their living and working environment for their quality of life and self-definition, and to the extent that their daily life includes an interaction with the non-human natural world, this environment and this interaction needs to be of a fair quality. This 'fair quality' should be defined by the institutions (democratically controlled) administering distributional justice. Background justice cannot be reduced to monetary terms. The equality of living and working conditions can also become undermined by the cumulative effect of the inadequacies of market rules or by historically unequal starting conditions. In this respect the plight of indigenous peoples, racial bias, gender bias and colonialism can give rise to background environmental injustice.

From the point of view of ecological justice it is clear that Rawls would view some environmentalist philosophies as examples of reasonable comprehensive moral doctrines. In Rawls's scheme of justice they would have to get along with other doctrines which take little account of the environment, or a different account of the human–nature relationship. Rawls's concern is to find institutions in which such values can find expression, not to insist that they become dominant. He uses his argument to support the institutions of democracy but he does not specify the inadequacies of current arrangements nor suggest modifications. Here we might want to extend Rawls and ask whether ecological values do in fact have a fair chance of entering debate. We have seen from the examples of Chapter 1 that they tend to occupy the margins, sometimes making incursions into current modes of co-ordination but only in an *ad hoc* and reactive way. The institutional means for arriving at co-operation are deficient in only allowing expression of ecological values in a minimal way and after decisions are taken when it is frequently too late.

We have suggested that both environmental and ecological injustice today have roots, not just within national systems but between them. Rawls substantially modifies his theory of justice for application to international relations (1993b), but we will postpone discussion of the wider question of the justice of transnational institutional structures until Chapter 7.

COMMUNITARIANISM

Whereas each of the former theories seeks to eliminate, or at least reduce the element of moral imperative, communitarianism brings it back, locating it in the authority of moral communities. Sandel (1982) takes up the question, implicit in Rawls's (1971) theory, of where social purpose comes from. He argues that within Rawls's theory can be found a variety of assumptions about 'community' as the source of values. These assumptions, he claims, are necessary to Rawls's arguments but enter as incoherent and isolated ideas. Rawls's conception of community is one which may be freely chosen by 'antecedent individuated selves'. It is not one which can enter into the creation of the self. Sandel draws a parallel between a society that is a 'community' and a 'just' society:

> a society ordered in a certain way, such that community describes its basic structure . . . constitutive of the shared self-understandings of the participants and embodied in their institutional arrangements, not simply an attribute of certain of the participants' plans of life.
>
> (Sandel, 1982: 173)

Community is thus viewed as the source of identity which in turn is the source of conceptions such as justice. In a world of communities, each defining justice in a different way, there can be no justice between communities. All that one can hope for, presumably, is a world in which the communities develop a way of living together that is not destructive of each other or of their shared planetary environment.

There is a contradiction between the emancipating project of deontology which seeks to liberate the person from the confines of culture, history and territory, and 'depth of moral character' (in Sandel's expression) gained by immersion in norms, ideals, places and a situated history. Sandel acknowl-edges this contradiction when he writes: 'As a self-interpreting being, I am able to reflect on my history and in this sense distance myself from it, but the distance is always precarious and provisional, the point of reflection never finally secured outside the history itself' (*ibid.*: 179). There may, however, be a real conflict within a community between those who want to enforce the community's norms and those who want to escape them. The triumph of the former can, as we have seen in many parts of the world in this century, bring extreme oppression. Commenting on Heidegger's ethics, Vogel observes:

> If each community's destiny is specific to its time and place, then one worries that the larger world is composed of a plurality of provincial communities each of which is closed in on itself with its own partic-ular moral self-understanding. Historicality does not account for how we move and mediate among different contexts, heritages, communi-ties, or how we adjudicate among conflicting interpretations of our own community's destiny. It offers no account of the sense that all other human beings share in 'our' destiny, and that honoring this requires listening to the perspectives of others from beyond the horizon of my or our prejudices, of suspending our projections for the sake of others who may have been excluded.
>
> (Vogel, 1994: 69)

Human authenticity is achieved on the 'border' between the community and the universal. Humanity is marked by the tension between sociality and the striving to transcend sociality.

> We coexist authentically and so form an 'authentic We' when each feels that he belongs to a common project yet encourages the others to pursue the project in a way that attests to their own individuality. This keeps the group from becoming a mere collectivity in which each must subordinate his own freedom in order to further shared ends. Every authentic 'We' is fragile because it is composed of self-responsible members who live at the boundary of this membership.
>
> (*ibid.*: 79)

This boundary area resonates with Krall's (1994) idea of the 'ecotone' which we discuss in Chapter 6: in this case the philosophical domain where two apparently contradictory possibilities or desiderata co-exist.

Most communitarians recognise, therefore, that the conditions for a 'self-interpreting being' to exist within a community demand the establishment of principles which transcend community. American communitarians like Bellah *et al.* (1985) and Etzioni (1993), in attempting to recover the occluded values of duty, care and responsibility in American society do not wish to abolish the common background of that society which espouses the principles of individualism and self-determination. There has to be a 'supra-community, a community of communities, – *the* American society' (Etzioni, 1993: 160). Charvet (1995) seeks to establish the (transcendant) 'principles of just cooperation' based on rights, markets, freedoms and property. So we return to the questions posed by Rawls concerning the distributional justice of a supracommunity, a 'union of social unions'.

As we have already argued (Chapter 2), communitarians have not articulated the institutional form of a 'community of communities'. They leave the field open to the existing co-ordinating mechanisms of capitalism and the market. Yet in our view it is precisely in this international and now global domain that the idea of community and public life is in most need of reinstatement. This reinstatement becomes the more urgent if new conceptions of the self inclusive of the non-human world are to enter the dialectic of justice.

DISCOURSE ETHICS

Public dialogue has always been a cornerstone of democracy. Dialogue is implicit in J.S. Mill's conception of democracy and in Rawls's 'justice as fairness'. Freedom of speech understood simply as the right to express an opinion 'into the void', as it were, trivialises communication. Public dialogue is about the freedom to engage in a meaningful two-way process. Benhabib sums up the place of public dialogue in democracy:

> In a democratic polity agreement among citizens generated through processes of public dialogue is central to the legitimacy of basic institutions. Such dialogues submit the rationale behind the major power arrangements of society to the test. Insight into the justice or injustice, fairness or unfairness of these arrangements gained as a result of such dialogic exchanges results in public knowledge won through public deliberation. . . . Perhaps the most valuable outcome of such authentic processes of public dialogue when compared to the mere exchange of information or the mere circulation of images is that, when and if they occur, such public conversations result in the cultivation of the faculty of judgement and the formation of an 'enlarged mentality'.
>
> (Benhabib, 1992: 121)

Benhabib reconsiders Kant's categorical imperative in the light of her ethic of judgement. What does it mean for all people to be treated as moral equals? It must mean that all people have a right to be treated as equals not only as *subjects* in (and subject to) the application of moral principles but also as *agents* in the making of such principles. In the idea of judgement Benhabib fuses the act of making with the act of applying principles. A judgement is a unique event in which a mixture of principles is applied in a specific situation. For each act of judgement a *different* multifacetted principle emerges. Thus, she writes:

The distinction between moral judgement and moral principles, between general rules which guide and govern our moral action and conduct and the specific form these rules assume in specific actions, events and situations helps us see how room may be made in Kantian theory for the exercise of moral judgement.

(ibid.: 136)

Kant's formula for *reflective* (as opposed to determinate) judgement based on universal communicability becomes the basis for Benhabib to reformulate Kant's categorical imperative from 'Act in such a way that the maxim of your actions can always be a universal law of nature' to: 'Act in such a way that the maxim of your actions takes into account the perspective of everyone else in such a way that you would be in a position to woo their consent' (*ibid.*).

This precept seems at first sight to place excessive weight on consensus. But Benhabib makes clear that this is not what she has in mind. Judgements have to be made between and among conflicting interests. The capacity for judgement is not that of empathy, 'for it does not mean emotionally assuming or accepting the point of view of the other. It means merely making present to oneself what the perspectives of others involved are or could be'. Each person is to be treated as one 'to whom I owe the moral respect to consider their standpoint' (*ibid.*: 136–7).

Like Rawls, Benhabib extends Kant but with the tools of Arendt. Although she is unwilling to return to decontextualised and abstract conceptions of 'the person' she none the less has to rely on some such idea to explain why contextualised persons owe each other respect. That persons must remain persons whatever their cultural context is something which has to be explained, since failure to do so licenses unmitigated violence (i.e. treatment of persons as things). Habermas and Apel look for an explanation, not in the fact of the person (though a Kantian view of the person is still implied), but in the fact of communication.

The central question for Habermas is: how do ethical positions constituted by different contexts find common ground in defining what exists and what is right? Habermas accepts that our understanding *both* of what exists and what is right is relativised to the context in which the activity of 'understanding' occurs. This context, as McCarthy (1984: xxiv) notes, is not just constituted by culture but by 'institutional orders and personality structures', and, one could add, by language games and regions – following Wittgenstein as Habermas does. The context is not just that of the public sphere as traditionally understood, but of everyday life. Habermas does not thematise the family as a site of political domination in the way that feminists have insisted it should be (see Fraser, 1987). But nor does he exclude it. In fact it would be highly inconsistent with Habermas's viewpoint to do so. Discourse ethics must apply at every site of communication both intimate and formal.

Habermas writes:

> Every process of reaching understanding takes place against the background of a culturally ingrained preunderstanding. This background knowledge remains unproblematic as a whole; only that part of the stock of knowledge that participants make use of and thematize at a given time is put to the test.
>
> (Habermas, 1984: 101)

Whatever is discussed is assigned to one of three 'worlds': 'A lifeworld forms the horizon of processes of reaching understanding in which participants agree to discuss something in the one objective world, in their common social world, or in a given subjective world' (*ibid*.: 131).

It is the process of reaching understanding itself (in whatever situation it occurs) that contains the germ of a political ethic. Enfolded in the very idea of 'reaching understanding' among real 'situated' people with real differences are two main ideas. First, reaching understanding is contrasted with and in opposition to strategic action aimed at winning, thus:

> Participants in argumentation have to presuppose in general that the structure of their communication, by virtue of features that can be described in purely formal terms, excludes all force – whether it arises from within the process of reaching understanding itself or influences it from outside – except the force of the better argument (and thus that it excludes, on their part, all motives except that of a cooperative search for the truth).
>
> (*ibid*.: 25)

Second, since claims to truth emerge from cultural contexts, the participant must be able to recognise the lineaments of that context and how it influences the arguments advanced. Thus speakers must speak authentically within context and acknowledge the validity of the context-laden arguments of others: Habermas posits that people are able to discover both what they are and what they really want and need through authentic interaction in conditions of equality of power, thus, in a sense, in the absence of power (see Low, 1991: 251).

Addressing the limits – among them ecological limits – of the efficacy of human political institutions based upon the nation state, Apel (1990) argues for 'a macroethic of co-responsibility'. The new institutions of the global economy, he says, have immense power to co-ordinate human behaviour, shrinking time and space to almost zero. They also force into our awareness 'the new relationship between man and nature, or rather between us and the part of nature that constitutes the human ecosphere' (*ibid*.: 26). We are individually expected to take responsibility for action which has distant effects in time and space both on other humans and non-human nature, but we have few ethical guidelines to regulate our conduct towards the distant other. The institution of the nation state seems unable to embody our responsibilities to the other. What is needed is a universally valid ethic for the whole of humanity which accepts and even guarantees the plurality of human conceptions of the good. That ethic, in Apel's view, is to be found in the practice of discourse.

We are faced, however, with the question of how a 'lifeworld' constituted by co-operative discursive relations between humans can penetrate the institutionalised human relations created by the rules we set up to co-ordinate human activity for purposes of production. Dryzek (1990) is impatient with the failure of Habermas to become specific on the implications of his theory for practical politics and institutional arrangements for democracy. Perhaps in order to avoid reification (and to avoid the problematic public–private distinction), Habermas takes system and lifeworld to be *aspects* of the same social world. However, practically speaking much human time goes into system-maintaining activity in which persons are treated and behave as so many ciphers: bearers of costs, wages, income, capital, age, gender, ethnicity and all the rest. This activity can be distinguished from activity in which people interact with one another and with non-human nature as persons. The central question is *how* values emerging from the latter kind of activity can be made to have some influence on the former.

Dryzek's answer lies in what he calls 'discursive designs'. Approaching modernity from the angle of institutional critique, Dryzek (1990; 1994) argues that the three forms of political institution evolved over the last two centuries are inadequate to deal rationally and ethically with today's distributional and ecological issues. These institutions are the capitalist market, the administrative state and liberal representative democracy (in which politically accountable officials – politicians – attempt to steer the administrative state). In combination, Dryzek thinks, these institutions are apt to compound rather than compensate for the errors inherent in each. The public sphere as traditionally understood is today dominated but not completely occupied by the rule structures sanctioned by these institutions. In the interstices of the power structure, however, 'public spheres' have increasingly formed for the examination of problems in which the conditions of ethical discourse become relevant.

These 'public spheres' are the more or less formalised meetings in many policy contexts to resolve disputes: mediation of civil, labour, international and environmental disputes (sometimes known as 'informal justice'), regulatory negotiation, and problem-solving workshops (Dryzek, 1990: 44, cites Wall, 1981; Harter, 1982; Gusman, 1981; Fisher and Ury, 1981; and Burton, 1979). Dryzek, in fact, seeks to redefine 'the public sphere' in a discourse-ethical way as, 'the space in which individuals enter into discourse that involves mutual respect, openness, scrutiny of their relationship with one another, the creation of truly public opinion, and, crucially, confrontation with state power' (Dryzek, 1994: 186). The reform of politics in a way which would favour ecological rationality requires the further growth at every level (internationally, at the boundaries of the state and within the state) of public spheres through discursive designs, and the evolution of universal ethical principles to govern their operation.

Discourse ethics moves us further away from the transcendental ethic of a universal rule towards the universal pragmatic ethic of political process. But there are some difficult matters to be resolved if discourse ethics are to have a practical impact on how humans govern their affairs for the purposes of environmental and ecological justice. At the interpersonal level, discourse ethics assumes that persons are able to free themselves to some extent from the contexts which pre-define their interests. While the experience of negotiation over the environment shows that this can occur, it has not been shown that it always or necessarily occurs (see Young, 1994). Moreover, discourse ethics seems to require a capacity to include the self-defined interests of others within the scope of the interests of self. The expanded sense of self advocated by ecologists is something we discuss in Chapter 6. However, what we have to consider is how an expanded sense of self may influence the world's governing institutions.

At the institutional level Dryzek may be too ready to dismiss the ecological effectiveness of the administrative state controlled by the full range of democratic institutions (including separation of powers, freedom of communication, information and assembly, majoritarian voting, the right to organise collectively and withdraw labour). As we shall see in the next chapter, the powers of a legitimate state have provided efficacious means for controlling and regulating environmentally damaging activities. While it is true that such regulation can work against environmental justice and the state can be co-opted to the side of 'growth'-oriented capitalism, there seems no other mechanism capable of regulating or transforming capitalism. 'Discursive designs' may play an important role in inserting wide-ranging social values into the 'system', but while capitalism as a system exists, the countervailing power of the administrative and regulatory state can hardly be dispensed with. At the international level we cannot be content with the evolution of incipient discursive designs which are nevertheless constrained by the rules and structures of global power. These rules themselves may need to be changed in order to facilitate the further growth of public spheres with the capacity to inject social values into co-ordinating systems. So consideration must be given to the institutional – rather than just political – means for changing the rules. We return to this question in Chapter 7.

Finally, new difficulties arise if, as Dryzek observes, 'discursive designs promote sensitivity to signs of disequilibrium in human–nature interactions because their *sine qua non* of extensive competent participation means that a wide variety of voices can be raised on behalf of a wide variety of concerns' (1994: 192). The acceptance of the voices of the non-human expressed through their human interpreters and advocates raises important questions

of priorities as between human and non-human nature and of communication between the two.

POLITICAL JUSTICE AND ENVIRONMENTAL CONFLICT

Until quite recently philosophical debates about justice have centred on the
attempt to establish claims for the supremacy of one or another system of
thought as universal and definitive. In philosophical debates, protagonists
and antagonists have examined each other's positions and exposed the weak-
nesses of each other's arguments. However, there is one point on which all
the participants in the debates are agreed: debate itself is the normal condition for the production of knowledge in human society. Moreover, while
Rawls's idea that the basic structure of a society should be arranged on the
basis of impartial judgement is laudable, none of the participants in the
debate can themselves claim such impartiality – including, as he now
acknowledges, Rawls himself. Debates about justice cannot be detached
from political debate in general, which in turn cannot be isolated from
power struggles in society. There is no Archimedean point outside the world
from which to change the world. The subtlety and complexity of human
experience cannot be enfolded within a single logical system. The achievement of justice is a dialectical project.

We have to think about how this dialectical project can best be pursued
in the world as it is currently structured. Let us once again revisit the practical circumstances of environmental conflict with which the book began.

The Brent Spar case demonstrates both the strength and weakness of
current international governance. The Oslo Convention became the framework under which state power could be mobilised to stop the loading on to
the environment of some of the costs of oil production. The dumping rights
of Shell came into conflict with the common-use rights to the North Sea
shared by the populations of several nations. The Oslo Convention is an
example of an incipient 'discursive design' existing between government,
state and market systems. Nevertheless, although there was public pressure
and argument, the issue of justice was not articulated. A reasoned judgement in the public sphere was not made. We cannot say that the events
surrounding the Brent Spar added anything to 'enlarged thought'.

The French nuclear testing in the Pacific also demonstrates the weakness
of international governance to control a matter of vital significance to the
future of humanity and the planet. The case demonstrates the continuing
strength of the sovereign nation state. At stake is the right of a nation to
distance itself from the environmental risks of its self-interested behaviour
and to impose that risk upon other nations. The economic needs of the
French (to support a viable nuclear industry) were in conflict with the environmental needs of the peoples of the Pacific. There is no transcendental rule
nor, in this case, a discursive institution under which the power of the nation

state could be checked. The expression of international outrage on the part of even such a formidable economic power as Japan proved of little effect. And since there was no general election in prospect, the French state was impervious to public opinion within its own territory even though majority opinion in France opposed the continued testing. Liberal democracy failed to control the administrative state. Australia, France's uranium supplier, refused to withdraw supply on the grounds that, in that event, other suppliers would quickly take Australia's place. The market has failed to prevent the production of deadly weapons even though few people desire it.

In the OK Tedi dispute the parties found an institutional means of opening the case for public debate. But a compromise was reached only after immense environmental damage had already been done. The court case hinged on the amount of compensation to be paid by the company to villagers who were not included in the original agreement to which the company was a party. The out-of-court settlement in effect recognised the right of the villagers not to have their environment damaged and the duty of the company to stop damaging it. That was the simple and narrow basis of the dispute. The right of a company to contract with the government of a poor nation state to extract minerals in an environmentally destructive manner was never in dispute. The justice of the rules of international trade which encourage such deals between multinational investors and governments was not an issue. Not considered were alternative and more sustainable ways of meeting Papua New Guinea's development needs. Not considered was the option of reducing consumption so that such mining became unnecessary (see, for example, the work of the Wuppertal Institute: Lovins *et al.*, 1997). The environmental rights and needs of the local villagers were not considered against the rights and needs of the final consumers of the products to which the mine contributed. Ecological justice, entering the damage to the local ecosystem as an instance of damage to the planet, could not be considered.

One may elicit many problematics from these cases. For example, a principle which emerges is that those who create environmental risk should bear the full burden of that risk. It does not seem fair in a global sense for the French government, merely because it 'owns' a few small Pacific atolls, to be able to locate the risk of nuclear testing so far from the territory of France. Ecological rationality suggests that those who produce environmental hazards should suffer from them. If consistently applied, such a principle would maximise feedback in such a way as to deter the production of risk. The principle, however, contains the assumption that agents producing the risk are territorially situated, that they occupy a place in the biosphere, and that the risk produced may be located in the place occupied by the agent. With persons, communities and states this is the case. They occupy a place. But it is not so with multi-national corporations. They are not territorially situated. Even if their workers and personnel inhabit a place, the corporation itself does not. It is a virtual entity. Unless and until the *virtual* reality of the corporation can be linked to the *actual* territorial reality of its workers via internal democracy, it will remain beyond space and largely (under the current world constitution) beyond control. Although headquartered in Australia, BHP cannot be made to feel the territorial impact of the risk it generates. Regulatory strategies are required which would be effective in the case of such non-place entities.

Conceptions of justice do not originate in the abstract logical systems of philosophers but in the everyday reflections of people upon their actions and the consequences of those actions, and the public discussion of ethical standards. Philosophers only later codify and try to make sense of these reflections and in so doing give us something further to reflect upon and act on. Bauman's (1995: 17) appeals for the 'confrontation of chaos' seem unduly hyperbolical and panicky. It is not chaos we confront but the vibrant diversity of our human praxis, some of which is concerned with finding universal standards for humanity, and some of which is embedded in and cannot be detached from its cultural context.

Under our present system corporations are entitled to the property they own. States are entitled to sovereignty over the territory they occupy. Both states and corporations act competitively to facilitate the accumulation of capital. But, as Kant ([c. 1798]1991: 158) observed, 'all nations stand originally in a community of land, though not of rightful community of possession and so of use of it and property in it; instead they stand in a community of possible physical interaction, that is in a thoroughgoing relation of each to all the others of offering to engage in commerce with any other'. Today we would say that all nations stand in a community of the biosphere. But the *inter-national* system makes it very difficult for anything but property rights and sovereignty to be given practical effect in conflicts over the environment.

Both contractarian and communitarian concepts are important markers in the dialectic of justice. Galtung (1994: 61) views the establishment of justice primarily in process terms, as the transmission – sending and receiving – of norms for the purpose of 'satisfying human needs through human rights'. He wants (at least) two channels to be open: one ultimately at the world level and one rooted in local cultures and networks. These he terms an 'alpha and a beta' channel, the former being the large, vertical and fragmented structures typical of bureaucracies and corporations, the latter being small structures, 'not necessarily quite horizontal but more *á l'hauteur de l'homme*'.

The primary issue in a justice ethic is not plurality or variety or incompatibility of ethical principles, but exclusion and inclusion. In our present world we have to consider what principles have real effect and what are excluded from consideration. We have to consider what institutions might be created which will embody a more inclusive consideration of justice. Such institutions will themselves have to rest on principles of justice.

99

CONCLUSION

We have seen in the first sections of this chapter that justice cannot be tied to a fixed and static formula. Justice emerges from a discursive struggle, a dialectic, which is entwined with politics and power in the material world. This dialectic finds expression, not only through the politics of pressure, but also through institutions of governance, some of which have themselves been moulded by the politics of pressure. Institutions channel both power struggles and discursive politics. The class struggle as it took place throughout this century presupposed the institutions of the nation state and the market. The nation state and its liberal discursive foundations were seen to be the target for democratisation in such a way as to allow the expression of the demand for justice as need. That expression led to forms of compact between opposing powers which instituted different forms of welfare state allocating functions between governments and markets. The discursive justification and critiques of states and social compacts has led to different societal theories of justice which take the nation state as the primary unit of governance: utilitarian, contractarian, entitlement, communitarian and discourse ethical.

In recent years institutions have developed which are in a real sense more powerful mechanisms of governance than the nation state. They are the global corporations and the various structures such as 'the financial market' which regulate their relations. These structures represent a global system of governance. Corporations are non-territorial entities of governance whose principal sanction is the right to exclude (dismissal, redundancy, downsizing or other euphemism), a right which most territorial democratic states have given up (banishment, transportation). The corporations also use the state's monopoly of legitimate violence to attain their economic ends where these ends cannot be attained by the threat of exclusion. Global governance operates on the basis of a particular and selective interpretation of justice: basically a mixture of utilitarianism and entitlement theory. Whatever alternative conceptions of justice lower-level institutions such as local governments and even some small corporations might be permitted to express and try to embody, the entitlement theory – property rights, the sanctity of commercial contract, sovereignty – today dominates the upper echelons of power to which all else is ultimately subordinate. This power is not completely determinant but it is formative. It exerts a continuous pressure upon all subordinate structures to adapt to its logic.

This institutional framework is the setting for the playing out of the ever more pressing contradiction between exploitation of the environment and its conservation. The unfolding of this contradiction presents the politics of the twenty-first century with challenges even greater than those of the twentieth. In so far as the opportunity for discursive politics exists, the attempt must be made once again to widen our understanding of justice so that a fuller range of interpretations can be inserted at the highest level of power.

This widening must include the distribution on a global scale of environmental quality; for the global level is where the upper echelons of power are now situated. As in the twentieth century, the political struggle will be shaped by the institutional framework and its discourse, but that framework is the global and not the national framework. The question is how can global institutions be shaped to express and embody wider considerations of justice?

The discourse of justice will also include nature as, in some sense, the subject of justice. But what can this mean? What is it for animals and rainforests to have merit? Can a sense of 'diachronic fairness' include fairness to the biosphere? Does such inclusion undermine the Kantian moral distinction between the treatment of things and the treatment of people? The human orientation to nature as the instrument of human satisfaction served in the development of institutions which exploit nature to the maximum for the production of commodities. If that system of production is now threatening to destroy the source of many kinds of human satisfactions, what does this mean for the evolution of discursive structures, moral principles, political institutions, and what new antinomies will now appear? These are the questions which we will have in mind in the chapters which follow in which we further interrogate the literature of environmental justice.

5
ENVIRONMENTAL JUSTICE
Distributing environmental quality

> If enlightened self interest is the principle of all morality, man's private interest must be made to coincide with the interest of humanity. . . . If man is shaped by environment, his environment must be made human.
>
> (Karl Marx and Friedrich Engels, 1845: 131, *The Holy Family*)

INTRODUCTION

In previous chapters we reviewed the major theories of justice and their conceptual elements. In this chapter we consider the meaning of justice for the distribution of environmental quality and risk. This distributive question is, of course, inescapably spatial, given both the materiality of nature and environmental diversity at local, regional and global scales. As will be explained, the rubric of 'environmental justice' has been inscribed in debates in the USA concerning fairness in the distribution of environmental wellbeing that have flourished in that country over the past decade. Our intention here is to assess environmental justice understood as a distributional precept against the different justice perspectives discussed in the preceding chapters: needs, rights, deserts, and political justice. We compare the work of those who have already discussed justice in the context of the environment with our own framework.

Making the environment 'human', as we shall see in Chapter 6, does not dispose of the question of ecological justice. A good environment for humans is not necessarily the same thing as a good environment for non-human nature. Yet Marx and Engels were referring to the evil of *humanly* created environments which were in every sense *inhumane*: the filth, squalor and overcrowding of the poor neighbourhoods of industrial cities. With both humans and non-human creatures, a *humane* environment is one in which their needs are met and in which they can optimally flourish.

Environmental quality is a central aspect of wellbeing for individuals and communities, and it is therefore a critical question for justice. Like any other dimension of wellbeing, environmental quality comprises both 'good' and 'bad' elements which are distributed across communities, nations and the globe. Obviously social values have an important role in determining both the nature of these distributions and our satisfaction with them. As we have already argued, the fact of divergent individual and communal values means that environmental justice cannot be seen as a simple, ahistorical ideal. Not only the quality of an environment, but the justice of its distribution may be evaluated in different ways.

So, we may have a land use for which there is an agreed level of social support, but whose physical proximity may actually reduce the environmental wellbeing of individuals. How, then, is the social and spatial distribution of such land uses to be agreed? Do all individuals and communities get a chance to express their values within decision-making frameworks that decide the distribution of environmental quality? Are there ways for the values of the lifeworld to enter the decision making system? What about the distribution of 'environmental goods' – those uses which are both socially valued and have the capacity to enhance individual wellbeing? In the absence of formal political mechanisms for ensuring fair allocations of 'good' and 'bad' land uses between individuals and communities, it appears that in western countries other social mechanisms, frequently centring on class and race relations, have acted as proxy distributors of environmental quality. We will examine in this chapter a variety of literature in the USA, Britain and Australia, which has shown how many local communities, often defined socially in terms of class or race, have accommodated an unfair share of environmentally injurious land uses.

Of course, as Ulrich Beck (1992; 1995) has shown us, the question of environmental 'quality' has increasingly dramatic implications in contemporary capitalist societies, and, for that matter, the globe. In the past, our political frameworks have valued the environment in instrumental terms, as a resource to be exploited for the production of use values which can then be distributed amongst communities and within humanity in general. But we are well aware now of the inadequacy of this ethical viewpoint and the disastrous environmental consequences of the industrial transformation of nature over the past two centuries. Beck has explained how capitalist modernity and its Promethean logic has produced potent industrial residuals which threaten human and non-human life at every geographical scale. This, in a sense, is our most dramatic injustice to nature – the production of environmental risks which now imperil the globe and all life within it. Moreover, these new hazardous substances, and the land uses associated with their production, storage and destruction, must be allocated socially and geographically, adding a new urgency to struggles for fairness in the distribution of environmental goods and bads. The distribution of such 'unwanted land uses' can, of course, occur at a variety of scales: between communities, cities, regions and nations. Indeed, the increasingly effective hostility of local communities in western countries towards hazardous waste facilities has encouraged an international trade which has sought to dump dangerous industrial by-products in developing nations. This 'traffic in risk' both imperils the wellbeing of the impoverished masses in developing countries, and also threatens to entrench the injustice of global uneven development.

This chapter is in two main parts. In the first we review the various theoretical and policy literatures which have addressed the distribution of environmental quality within western countries, notably the USA. The second part considers the environmental justice question at the supranational level, focusing on the international trade in risk and the implications of this for uneven development.

JUSTICE WITHIN NATIONAL ENVIRONMENTS

Both environmental quality and environmental values are distributed at a variety of spatial scales, ranging from the local–communal level, through regions and nation states, to the entire globe. These distributions, which are highly variegated in socio-cultural and spatial terms, interact to produce a diverse and shifting landscape of ecological politics. This fact confounds attempts to arrive at universal prescriptions of what is a fair distribution of environmental quality for any scale of analysis. This, then, also complicates the political task of allocating land uses and activities which impinge heavily on environmental quality.

However, like Beck (1995: 75–6), we are not prepared to endorse value relativism either as a virtue or as an inescapable fact in the 'age of risk'. There are *objective* dangers arising from contemporary industrialism, in the form of toxic wastes and other hazards, which cannot be socially distributed merely through a system of culturally derived preferences. The danger is not just a matter of opinion. Too often cultural relativism is a mask for anti-democratic politics and even localised tyranny. Even where some form of democracy can be assumed for all social contexts, it is doubtful that all communities will possess 'perfect information' concerning the nature of the environmental risks they may be asked to carry in the form of hazardous land uses. As will be shown in this section, a collusion between markets and racially discriminatory anti-ecological local politics has produced a racialised pattern of risk in the United States, meaning that many urban coloured communities now bear a disproportionate share of the environmental risks that arise from that nation's hazardous industries.

At present the regulation of the distribution of environmental quality falls on national governments, and their subsidiary (regional and local) states. In this section we will examine how the relationship between states, markets and local communities determines national distributions of environmental quality. This will clarify some of the institutional issues for the following section which considers environmental justice at the supranational level – an environmental policy arena marked by the absence of a regulatory state.

Socio-spatial justice

Geographers have become aware that justice must be realised (or violated) within concrete environmental settings. From the late 1960s, a range of social scientists, especially urban geographers, have applied justice-related concepts to the analysis of spatial patterns in western countries. Hay (1996) argues that the ethical notions of equity, fairness and justice have been used interchangeably by spatial social scientists as the evaluative bases for geographic measures of wellbeing. As Hay (1996) explains, other ethical concepts, notably procedural justice and desert, have proved harder to operationalise spatially. None the less it has been recognised by some geographers (e.g. Blomley, 1985; 1989) that both spatial and temporal consistency in the application of the law are preconditions for juridical impartiality (the principle of the 'rule of law' elucidated, for example, by Hayek).

In the main, spatial social science has followed the utilitarian practice of measurement of distributional outcomes. Reflecting this substantive concern, the notion of 'territorial justice' emerged as one early socio-spatial measure of fairness (e.g. Davies, 1968; Harvey, 1973). Davies (1968: 39) defined territorial justice as: 'an area distribution of provision of services such that each area's standard is proportional to the total needs for the service of its population'. Although Johnston *et al.* (1994: 300) define geographical justice as 'the empirical and theoretical study of the . . . fairness of the geographical apportionment of benefits', it is true that much of the work undertaken beneath this rubric has focused on the distribution of publicly provided 'goods' (Boyne and Powell, 1991). The territorial justice principle was largely applied to analyses of the national and regional distributions of social services within western countries, notably the UK and the USA (e.g. Davies, 1968; Pinch, 1979; 1985; Curtis, 1989). For analyses of mainland Europe, see Mingione and Morlicchio (1993) and Petmesidou and Tsoulovis (1994). While this form of analysis was able to expose many discriminatory patterns of service provision, little attempt was made to relate such findings to the geographic distribution of social needs (Hay, 1996).

From the early 1970s, an important strand in geographical analysis measured spatial wellbeing as a person's relative position in terms of both accessibility to valued public services and proximity to undesirable land uses (Dicken and Lloyd, 1981). By the 1980s a voluminous literature had developed describing the relationship between the residential structures of major western cities (including race and class characteristics) and patterns of accessibility to land uses, especially 'salutary facilities', which were held to be socially valued (see, for example, Knox, 1982; 1995). Similarly, other geographic analyses focused on the locational patterns of 'noxious facilities', those which generate negative externalities for surrounding communities. Interestingly, environmental 'nuisances' were defined broadly, including such diverse land uses as airports, polluting industrial plants, football stadiums and community facilities for deinstitutionalised people (Dicken and Lloyd, 1981; Dear and Taylor, 1982; Harvey, 1973). Harvey (1972: 27–8) argued that the environmental quality of any land use depends upon the potentially diverse ecological and social values of its 'host' community. But this relativist argument is problematic as we shall see.

The welfare geography of Smith (e.g. 1975; 1977; 1979) and Knox (e.g. 1975) extended the territorial justice notion to include the distribution of a broad range of benefits, including public services and a range of other social 'goods' and 'bads'. Smith's considerable empirical investigations of wellbeing in the USA, Britain and South Africa measured geographic variation on a range of composite social indicators. Significantly, these indicators included a number of environmental variables, such as air pollution (see Smith, 1975) and built-environment quality (see Smith, 1977). Although avowedly norma-tive (like its disciplinary equivalent in economics), welfare geography was an essentially descriptive exercise which made scant reference to the socio-structural causes of spatial inequity. It thus had little explanatory power and limited political salience. (This is to echo Young's, 1990, general criticisms of distributional justice.) None the less, by demonstrating the highly spatialised distribution of social wellbeing, the welfare perspective foreshadowed the potential for a geographic analysis of environmental justice.

Other geographers, notably Badcock (e.g. 1984), Harvey (e.g. 1973; 1981; 1982; 1996) and Smith (e.g. 1984), sought to explain territorial injustice in capitalist societies through resort to political economy (especially Marxian social theory). The difficult ethical questions of justice were set aside in favour of structural explanations focusing mainly on unequal power. Harvey's earlier work helped establish a range of social scientific analyses which has sought to describe and explain the racial and class inequality in the distribution of well-being, most acutely represented in studies of the North American 'ghetto'. Both Badcock (1984) and Harvey (1973) have characterised the capitalist city as a 'resource distributing mechanism', highlighting how economic structures (e.g. relations of production) and institutionalised power (e.g. state forms) are inscribed in the urban form through differentiated patterns of wellbeing. The urban political economy approach demonstrated the capacity of the capitalist urban system to thwart the redistributive objectives desired by welfarist policy (Badcock, 1984). As a spatial concentration of market mechanisms and social power structures, the city was an unmistakable revelation of the tendency of capitalism to distribute socio-economic and environmental resources unevenly. Related analyses pointed to the ways in which social power was articulated in urban communal struggles over the distribution of environmental and economic resources (e.g. Janelle and Millward, 1976; Castells, 1979; Cox, 1973; 1979; Walker, 1981). Several commentators (e.g. Dear 1977; Johnston, 1984; Plotkin, 1987; Reynolds and Honey, 1978), for example, have observed how certain state institutional mechanisms, especially planning regulations, are used by the privileged classes to keep noxious land uses away from the places they occupy and concentrate such uses in poor and working-class neighbourhoods.

During the 1980s, the emphasis on justice in geography diminished as many critical analysts turned to socio-structural explanations of inequality, including variants of Marxism which were hostile to ethics (Johnston *et al.*, 1994). (Pirie's (1983) thoughtful analysis attempted to apply a socio-spatial ontology to justice, but his project seems to have gone no further.) Other theoretical currents, notably post-modernism and neo-liberalism, also problematised universalist ethical notions in the social sciences generally. By the 1990s, several geographers had begun to re-assert the importance of justice in socio-spatial analyses of wellbeing (Gleeson, 1996). Amongst these, Harvey (1992; 1993b; 1996) and Smith (1994) made notable interventions which sought to reinstate the importance of the justice principle in an academy increasingly dominated by relativist ethics. Both analysts have also pointed to the political saliency of justice in an era when market relations and neo-liberal politics dominate the globe. In a further recent development, a growing number of North American geographers (e.g. Lake, 1996; Lake and Disch, 1992; Pulido, 1994; 1996; Seager, 1993) have turned their attention to the issue of 'environmental justice' which has emerged from grassroots campaigns in the 1970s to now become a key focus of national political debates and federal policy in the USA (see, for example, the 1996 special 'environmental justice' issue of the journal *Antipode* 28(2) in 1996).

Environmental racism

A growing debate in the United States about environmental justice had its origins in the grassroots struggles of local communities during the 1970s against 'environmental racism' (Alston, 1990; Harvey, 1996; Sarokin and Schulkin, 1994). These struggles, involving both local communities of colour and a range of progressive groupings (notably churches and civil rights organisations), sought to oppose the racially discriminatory distribution of hazardous wastes and polluting industries in the United States. A range of minority groups were involved in these campaigns, including urban African-American and Latino communities and native American peoples residing on traditional lands (much of which had been poisoned by military and industrial uses). Importantly, this grassroots campaign emerged outside, even at times in opposition to, the mainstream of the environmental movement in the USA (Hofrichter, 1993). Activists pointed out that the environmental movement had concentrated on the ecological concerns of white, middle-class Americans, and had failed to identify and oppose the disproportionate burden of toxic contamination on minority communities. Hofrichter (1993: 2) attributes the toxic burden on communities of colour to 'the unregulated, often racist, activities of major corporations who target them for high technology industries, incinerators and waste'.

A seminal moment in the environmental racism campaign was provided in 1982 by the vigorous protests against the siting of a PCB landfill in a black community within Warren County, North Carolina (Cutter, 1995; Mohai and Bryant, 1992). The Warren County action saw prominent national civil rights leaders uniting with the local community in a campaign of civil disobedience (resulting in 500 arrests) reminiscent of the racial justice struggles of the 1960s (Goldman, 1996; Heiman, 1996). Shortly afterwards, a federal government study found evidence of racial discrimination in the location of commercial toxic waste landfills in one region of the USA (United States General Accounting Office (USGAO), 1983). Following this in 1987, the influential United Church of Christ (UCC) report on toxic waste patterns demonstrated that race was the central determining factor in the distribution of chemical hazard exposure in the United States (United Church of Christ Commission for Racial Justice, 1987). The broad findings of the landmark UCC report, were confirmed by later social scientific studies (e.g. Adeola, 1994; Bryant and Mohai, 1992; Bullard, 1990a; 1990b; 1992a; 1992b; Bullard and Wright, 1990; Mohai and Bryant, 1992), although in recent years a considerable number of analyses (e.g. Been, 1993; 1994; Boerner and Lambert, 1995) have also sought to 'debunk' the environmental racism thesis on methodological grounds. However, both Goldman (1996) and Heiman (1996) point out that many of these sceptical studies have been funded by risk-producing and waste-management industries.

By the early 1990s, several thousand groups had emerged to oppose inequitable distributions of land uses which threatened the environmental health of local communities (Bullard, 1993a,b,c). In many instances, community action was successful in either preventing the establishment of polluting facilities or ameliorating their effects through both voluntary and enforced agreements on site conditions. Also, the 1990s have seen the environmental racism movement refocus its political–ethical ideals around the broader notion of 'environmental justice' (Cutter, 1995). In 1991, more than 650 activists from over 300 local grassroots groups attended the First National People of Colour Environmental Leadership Summit in Washington, DC (Goldman, 1996). The summit adopted seventeen principles of environmental justice which extend the movement's focus on race to include other concerns, such as class and non-human species (see Figure 5.1). Cutter (1995: 113) argues that the movement has now transcended, without abandoning, its concern with communities of colour to 'include others (regardless of race or ethnicity) who are deprived of their environmental rights, such as women, children and the poor' – a definition endorsed by Hofrichter (1993).

WE, THE PEOPLE OF COLOR, gathered together at this multinational People of Color Environmental Leadership Summit, to begin to build a national and international movement of all peoples of color to fight the destruction and taking of our lands and communities, do hereby re-establish our spiritual interdependence to the sacredness of our Mother Earth; to respect and celebrate each of our cultures, languages and beliefs about the natural world and our roles in healing ourselves; to insure environmental justice; to promote economic alternatives which would contribute to the development of environmentally safe livelihoods; and, to secure our political, economic and cultural liberation that has been denied for over 500 years of colonization and oppression, resulting in the poisoning of our communities and land and the genocide of our peoples, do affirm and adopt these Principles of Environmental Justice:

1   Environmental Justice affirms the sacredness of Mother Earth, ecological unity and the interdependence of all species, and the right to be free from ecological destruction.

2   Environmental Justice demands that public policy be based on mutual respect and justice for all peoples, free from any form of discrimination or bias.

3   Environmental Justice mandates the right to ethical, balanced and responsible uses of land and renewable resources in the interest of a sustainable planet for humans and other living things.

4   Environmental Justice calls for universal protection from nuclear testing, extraction, production and disposal of toxic/hazardous wastes and poisons and nuclear testing that threaten the fundamental right to clean air, land, water, and food.

5   Environmental Justice affirms the fundamental right to political, economic, cultural and environmental self-determination of all peoples.

6   Environmental Justice demands the cessation of the production of all toxins, hazardous wastes, and radioactive materials, and that all past and current producers be held strictly accountable to the people for detoxification and the containment at the point of production.

7   Environmental Justice demands the right to participate as equal partners at every level of decision-making, including needs assessment, planning, implementation, enforcement and evaluation.

8   Environmental Justice affirms the right of all workers to a safe and healthy work environment without being forced to choose between an unsafe livelihood and unemployment. It also affirms the right of those who work at home to be free from environmental hazards.

9   Environmental Justice protects the right of victims of environmental injustice to receive full compensation and reparations for damages as well as quality health care.

10   Environmental Justice considers governmental acts of environmental injustice a violation of international law, the Universal Declaration On Human Rights, and the United Nations Convention on Genocide.

11   Environmental Justice must recognize a special legal and natural relationship of Native Peoples to the U.S. government through treaties, agreements, compacts, and covenants affirming sovereignty and self-determination.

12   Environmental Justice affirms the need for urban and rural ecological policies to clean up and rebuild our cities and rural areas in balance with nature, honoring the cultural integrity of all our communities, and providing fair access for all to the full range of resources.

13   Environmental Justice calls for the strict enforcement of principles of informed consent, and a halt to the testing of experimental reproductive and medical procedures and vaccinations on people of color.

14   Environmental Justice opposes the destructive operations of multi-national corporations.

15   Environmental Justice opposes military occupation, repression and exploitation of lands, peoples and cultures, and other life forms.

16   Environmental Justice calls for the education of present and future generations which emphasizes social and environmental issues, based on our experience and an appreciation of our diverse cultural perspectives.

17   Environmental Justice requires that we, as individuals, make personal and consumer choices to consume as little of Mother Earth's resources and to produce as little waste as possible; and make the conscious decision to challenge and reprioritize our lifestyles to insure the health of the natural world for present and future generations.

Figure 5.1 Principles of environmental justice adopted by the First National People of Color Environmental Leadership Summit
Source: United Church of Christ Commission for Racial Justice, 1991

The environmental justice movement

This broadening of political purpose described above has also extended the social and institutional reach of the environmental justice movement, which has 'moved from street-level protests to federal commissions, corporate strategies, and academic conferences' (Goldman, 1996: 131). Indeed, 'Environmental justice concerns are being institutionalized within government, academia and business, mediated in important ways by the press' (*ibid.*). Not surprisingly then, the emergent environmental justice movement attracted regional and national political attention during the late 1980s and early 1990s. Goldman reports that by 1996: 'Numerous federal, state, and local bills have been introduced to address various aspects of environmental injustice, addressing fair siting, citizen participation, compensation, and health research' (1996: 127).

In 1992, the US Environmental Protection Agency established an Office of Environmental Equity and published a report on the national distribution of ecological risks (USEPA, 1992). The movement's official recognition reached its zenith in February, 1994, when President Clinton signed Executive Order 12898, which required that every federal agency consider the effects of its own policies and programmes on the health and environmental wellbeing of minority communities (Cutter, 1995; Goldman, 1996). To date the achievements of the federal environmental justice programme have been modest, but certainly worthwhile, including remediation works at a number of contaminated sites, the improvement of some community health services, the funding of education and training campaigns in hazard awareness and monitoring, and the targeting of minority business enterprises in the awarding of EPA contracts (USEPA, 1995).

Despite its institutional successes, there are growing indications that the environmental justice movement may have reached its political high tide mark, at least for the foreseeable future. Goldman (1996: 126) argues that the movement 'may be entering the most difficult phase of its early history' as political opposition to environmentalism strengthens in the federal and state legislatures. The so-called 'Wise Use' campaign has united many industry and resource user groups who argue that US environmental standards and regulations are too stringent and represent an 'unjust' circumscription of private property rights (Helvarg, 1994; Harvey, 1996). The Wise Use lobbies have been successful in shaping the environmental and resource policies of the Republican Party which now controls Capitol Hill and many state legislatures (Helvarg, 1995). Generally speaking, the contemporary Republican political agenda calls for reductions to both funding and programmatic support for a range of state functions (Gillespie and Schellhas, 1994). Critically, these targeted functions include affirmative action programmes, redistributive social policies and environmental regulation, all key public policy elements for the environmental justice movement (Goldman, 1996; Heiman, 1996; Hofrichter, 1993; Sarokin and Schulkin, 1994). During 1994–5, the reality of the Wise Use threat to environmental justice was underscored when the Republican Party proposed radical funding cuts to the federal EPA. Green lobbies and most Democrats countered that these cuts, if realised, would undermine the principal federal institutional base of the environmental justice movement.

A further threat has emerged in the form of industry-sponsored research and legal manoeuvres which have sought to oppose the claims and activities of the environmental justice movement. As Goldman (1996: 132) notes, polluting industries and their allies in waste management have engaged a range of 'expert' commentators in order to deflect the political arguments of environmental justice activists with legal and scientific complexities: 'Now the academic guns have been loaded to defend the turf of expertise, raise the threshold of entry into the debate, and ensure that the burden of proof remains squarely on the backs of the victims of pollution.' Here we have a struggle to contain the impact of lifeworld values and contain the environmental problematic within the administrative state and its professional ancilliaries.

The environmental justice movement also faces a number of internally generated challenges and threats. There is a need for the movement to better define its politico-ethical purpose: many commentators and activists now argue that its established focus on the *distribution* of environmental wellbeing has actually entrenched the political power of polluting industry and waste management corporations. As both Heiman (1996) and Cutter (1995) have noted, the movement has tended to pursue 'environmental equity', meaning the equitable distribution of negative externalities. Heiman argues that the realisation of this political goal will hardly trouble polluting industry, which, after adjusting its locational prerogatives, will resume its risk generating production, only with the EPA assurance of fair equality of opportunity to pollute and be polluted (Heiman, 1996: 114).

Overlaying the two Rawlsian principles of fair equality of opportunity and the difference principle on an existing production system is going to change little. Rawls might say that 'pollution should be distributed in such a way that it is of greatest benefit to the least advantaged'. However, if we were to apply a Rawlsian approach, which included both distribution and production in a single calculus, the outcome might be different. Suppose that the producers and recipients of risk were to sit down together to decide on the production of LULUs (locally unwanted land uses) – a discursive design might, in fact, produce just such a situation. Applying a moderate veil of uncertainty, if not ignorance, the decision makers would not know about the *location* of the LULU whose production was being decided upon or the risks attached to it (not an altogether unlikely scenario given today's scepticism about scientific knowledge). Let us suppose that the LULU is, for example, a chemical plant. Might it not be that the decision makers would weigh seriously the option of doing without the LULU? The argument becomes one about 'odds' and the propensity of the decision makers to gamble with their backyards and their lives. Rawls might object that the veil of igno-rance is meant to apply only to the basic structure of society. A Kantian would say, however, that the contract is meant as a daily reminder to decision makers of their obligation to make only laws that can apply equally to all. Both a Rawlsian and Kantian argument could be employed within a discursive design to insist, on grounds of justice, that producers not only 'pay' but also suffer the risk of the hazards they cause through production.

The environmental equity approach has been reflected in variants of 'green' legal philosophy which have stressed the importance of procedural fairness to the resolution of ecological conflicts and the achievement of social harmony (e.g. Hoban and Brooks, 1987; Mandelker, 1981). Hoban and Brooks (1987: 219), for example, insist that, 'the equitable application of law and legal principles to . . . environmental problems is *the one indispensable requirement* if we are to clean up our environment and arrive at an environ-mentally just society' (emphasis added). Lake (1996) and Heiman (1996: 116), however, oppose the 'environmental equity' ideal, both for its naive faith in procedural justice in social conflict settings and the inability of distributional notions of fairness to problematise the structural and institu-tional sources of injustice (cf. Young, 1992; Harvey, 1996). Heiman (1996: 114) proposes that: 'environmental justice demands more than mere exposure equity . . . it must incorporate democratic participation in the production decision itself'. In this he is approaching something closer to Dryzek's 'discursive designs'. Indeed Dryzek (1994: 194) applauds the expansion of discursive democratisation within the economic sphere: 'economic organization should fall under discursively democratic political control'. As we argued earlier, democratisation of the corporation is necessary to reconnect the corporation to the spatial reality of the real world – 'real' as opposed to the virtual environment to which corporations increasingly retreat.

Heiman (1996: 119) further elaborates the environmental justice ideal as centring on 'community empowerment and access to the resources necessary for an active role in decisions affecting people's lives'. In Chapter 3 we saw that 'need' has been defined in just this way. The idea of empowerment derives its moral force ultimately from the Kantian idea of the person, and therefore the right to autonomy, which infuses conceptions of rights and needs and the virtue of citizenship (see Doyal and Gough, 1991; Bookchin, 1990; 1995a).

An additional internal problem for the environmental movement exists in the very scale of its considerable political success to date. The vast socio-political reach of the contemporary movement is evidenced by the profusion of grassroots activist groups and information dissemination networks (including a well-supported internet web site, 'EcoNet'). However, without a clear-sighted understanding of the meanings of 'justice' there is a danger that the 'mainstreaming' of opposition to environmental risks will further worsen racial and class disparities. To adapt Marx's observation, where rights conflict as they inevitably do, power can all too easily decide:

> As more communities try to block sites and prevent pollution in their backyards, those with the least political and economic power will be left with an even greater share of the toxic residues from our modern society. . . . As manufacturers downsize and consolidate their facilities, the plants posing the greatest potential hazards are likely to be left in communities that fit a particular demographic profile.
>
> (Goldman, 1996: 128)

At present, the environmental justice policy landscape seems dominated by debates over the 'fair share' allocation of LULUs among communities, variously defined. It seems that the environmental justice movement is in danger of being overtaken by the increasingly general awareness of, and antipathy for, hazard-producing land uses amongst the broader population, drawing the movement's critical energies into the quicksand of distributional politics. The answer, however, lies not in avoiding the question of conflicting values but resiting the discursive struggle from the distributional periphery to the production centre, and from the local to the transnational scale. In the next sub-section we explore the potential difficulties posed by this present 'environmental equity' policy focus, both for disadvantaged communities and the broader environmental justice movement itself.

DISTRIBUTIONAL POLITICS – UTILITARIAN SOLUTIONS

A critical thread of the rising popular environmental consciousness in the USA has been the growing awareness of the health risks posed by hazardous industries and waste management activities. Since the 1970s, there has been increasing opposition across most local and regional jurisdictions towards the establishment, and continued operation, of polluting and waste management activities (LULUs). The acronym 'NIMBY' (Not-In-My-Backyard) has been coined to describe popular antipathy for residential proximity to LULUs (Dear, 1992; Popper, 1981). The US planning and environmental policy realms have thus for some time now been overshadowed, even in some parts momentarily paralysed, by the LULU problem (Popper, 1981; 1992), which is seen in technocratic terms as an attitudinal paradox between social support for polluting industries (and their products) and local hostility towards land uses which host such activities (Dear, 1992). 'Social support', it should be noted, tends to be deduced from the absence of alternatives. No-one has actually bothered to test what the public wants in the light of debate and full consideration of costed options. The issue has been particularly acute within urban and regional planning frameworks, given that many local communities have successfully used zoning ordinances to exclude LULUs from their political jurisdictions.

Interestingly, the category, 'LULU', includes not only industrial activities, but also residential land uses for 'socially undesirable' people, such as homes for ex-prisoners, deinstitutionalised mental patients, people with AIDS and the like (Dear, 1992; Gleeson and Memon, 1994). Everywhere it seems that residential communities are fearing the 'contaminating touch' of land uses which produce unpleasant and risky side effects, be these chemical poisoning, nuclear radiation, physical assault or just the distasteful sight of modernity's human refuse loitering on street corners and in neighbourhood parks.

Technocrats have decried the proliferation of NIMBY opposition as a threat to US industrial and infrastructure development, which ultimately imperils national wellbeing. One can understand the frustration of technocrats and industrialists, summarised thus: 'Our fickle people want the goods we make, the comforts we supply, but refuse to acknowledge that these things must be produced somewhere!' Moreover, they complain, if new industries can't be established, jobs will be lost to overseas competitors and economic wellbeing will decline, especially for the industrial classes; if waste management and destruction facilities cannot be sited, then the orderly control of industrial residues – the ever-growing lake of toxic contaminants, the continually rising mountain of waste materials – will break down and national health will be jeopardised. Even Vice-President Gore has been moved to observe on the NIMBY problem:

> I have always been struck by the way a proposal for an incinerator or a landfill mobilizes a lot of people who do not want the offending entity near them. In the midst of such a controversy, no one seems to care much about the economy or the unemployment rate; the only thing that matters is protecting their backyard.
>
> (Gore, 1993: 355)

Thus, an ever-expanding scientific and policy literature has proliferated to advise governments and corporations on devising new locational strategies which will both identify the safest (and acceptably efficient) locations for LULUs, while allaying community concerns through the latest, most sensitive consultation methods (e.g. Carver and Openshaw, 1992; Gregg *et al.*, 1988; Lober, 1995; Massam, 1980; 1993). Thus, technocratic utilitarianism is becoming allied (in a highly contradictory relationship) with discursive design. The erroneous assumption is that there is a single welfare function which can be found with the use of technology and the right kind of discourse (e.g. geographical information systems (GISs), which aim to derive optimal site choices for LULUs aided by participation 'techniques').

Parallel to this literature, though emerging from more humane, liberal instincts, has been a set of policy-scientific discourses which have argued for 'fair share' distributions of LULUs across local and regional jurisdictions (e.g. Dear *et al.*, 1994). These discourses have noted the already inequitable geographical distribution of LULUs – from the racist locational patterns of polluting industry to the ghettoisation of facilities for 'social outcasts', usually in poor inner-city neighbourhoods – and have thus argued for corrective policy mechanisms, usually in the form of planning controls, which will ensure a more uniform sharing of the LULU burden. It is not difficult to see how the equity focus pursued by environmental justice activists has been submerged under a rising political tide favouring fair-share zoning and equitable corporate siting policies.

In some instances fair share policies have actually become law. In 1991, for example, the City of New York introduced 'fair share criteria' into planning regulations in an 'attempt to foster an equitable distribution of public facilities throughout the city' (New York Department of City Planning, 1991: 1). Most private agencies, however, were not bound by the criteria, which, in any case, used 'exhortatory, rather than mandatory, rules' to achieve their objectives (Valetta, 1993: 20). A 1993 review of New York's fair share criteria found that public agencies had not administered the new system effectively, and that local communities were still excluded from many critical facility location decisions (Manhattan Borough Board, 1993; see also Rose, 1993). Recently, the US Supreme Court ruled that planning regulations could not be used to exclude group homes for disabled people from residential areas (*AAMR News and Notes*, May/June, 1995: 1). The court ruling, of course, did not extend to the many other types of LULU, especially polluting industries.

Thus, while some form of 'environmental equity' may now have considerable moral–political force for LULU operators, it seems highly doubtful that this will be realised through fair-share planning regulations which aim to establish geographic uniformity in the distribution of risk. Political resistance from capital and the existing environmentally privileged classes to the institutionalisation of such 'territorial justice' would be immense.

Increasingly it seems that 'environmental equity' may mean simply that poor and coloured communities would be compensated for hosting risky facilities, both by polluting firms and wealthier social strata. Already a considerable literature has placed itself at the service of exasperated local and regional states, suggesting systems for the open allocation of LULUs to jurisdictions which ensure both procedural fairness and the compensation of those 'loser' communities which emerge from these processes with a disproportionate share of risky facilities. Armour (1991), Boerner and Lambert (1995) and O'Hare *et al.* (1983), for example, argue for compensation (taking both fiscal and non-monetary forms) of communities which host LULUs on the grounds of distributional equity.

Much of this literature, especially the contributions from economists, is founded on utilitarian notions of political justice (Hartley, 1995). Especially prevalent is the welfare–utilitarian assumption that all human preferences for social 'goods' and 'bads' can be measured in money terms (principally as price signals emerging through exchange mechanisms) and can therefore be equated, substituted, and even traded, for distributional purposes. Thus it follows that a common level of utility can be achieved across jurisdictions faced with the collective task of distributing the benefits and burdens that flow from a regional allocation of social and environmental resources. It is assumed that if communities are left to determine their relative tastes for various goods and bads, thus formulating unique 'welfare functions', it follows that a supervising body need only ensure the material equilibration of these local preference sets through monetary transfers (i.e. compensation) in order for distributional justice to occur. Thus O'Sullivan (1993), for example, suggests that the LULU conundrum can be solved by the establishment of facility auctions, supervised by a regional body with the authority to distribute a set of noxious land uses between local jurisdictions. The auction would involve each jurisdiction lodging monetary bids to prevent the location of LULUs within its geographic bounds. The lowest bidder would be required to take the LULU(s) but would be awarded the sum of the other bids as compensation. This process would serve both to monetise the composite 'environmental–economic preferences' of each community, while also providing a means for allocating, socially and spatially, the disutility represented by the LULU(s). From these two consequences, a uniform distribution of utility can subsequently be achieved through compensating monetary transfers. On the face of it, the auction model seems to recommend itself on the grounds of efficiency and equity. Or does it?

Gleeson (1995) has raised a number of specific objections to the various utilitarian models of LULU allocation, most of which rely on some form of compensation as a means of achieving *ex post* distributional equity between communities. Certainly the specious utilitarian assumption that different aspects of wellbeing, notably environmental health and economic security, can be first measured and then made commensurable through the medium of money ignores the kind of critique advanced by Self (1970), Rawls (1982), Elster (1982) and Jacobs (1991), as discussed in Chapter 4, that preferences are conditioned by social contexts and contexts are infused with unequal power. Monetisation of risk allows structural inequalities to be exploited by risk producers. Bullard (1992a) has termed this form of exploitation 'environmental blackmail', meaning the political–economic pressure to host a polluting industry which is frequently applied by firms and governments to poor and coloured communities in the United States. The price extorted, literally some quantum of public health, is hardly commensurable with the rewards of some measure of 'growth' (meaning business activity) in the long run. Even if monetisation made sense (which it does not) the fair price, all things being equal (which they are not), would be that which it would take to make the residents of the community freely move away from the source of pollution. That would be the fair price of the social and physical *environment* to those residents (assuming that 'the environment' is much more than the sum of a few plots of land). Since this never happens, it is also important to note that compensation is frequently a one-off event which does not recompense succeeding inhabitants of communities which accept LULUs in return for monetary (or other) rewards. Compensation may therefore entrench intergenerational inequity.

There are many other dangers in the utilitarian solution to the LULU problem. It certainly encourages LULU operators to target communities which are vulnerable to compensation, namely, in the USA, those poor and coloured neighbourhoods which already host a disproportionate burden of risk-producing activities. The strategy thus threatens to further ghettoise the geographic and social distribution of risk, raising the spectres of cumulative impacts and the environmental 'death' of places and regions. Moreover, the utilitarian view rests on the neo-classical idea of perfect information as the basis for fair and efficient exchanges: one may debate whether such a precondition is ever a feature of human reality, but it certainly does not apply in the complex socio-political contexts where LULU decisions are made (and sometimes imposed through various means). The popular understanding of industrial risks may be rising, but it is yet to converge with science's partial and hardly uniform opinions on the nature of hazards. Moreover, popular understanding of risks differs greatly between individual communities, and especially among the social classes with their varying access to scientific knowledge and official information. Heiman (1996) and Goldman (1996) argue, for example, that the environmental justice movement may well awaken the 'risk consciousness' of the middle classes, thus enhancing the power of wealthier communities to prevent the redistribution of LULUs into their jurisdictions.

The utilitarian approach has encouraged a 'consensual community' view in public planning and environmental policy realms in the United States. Critically, there is no appreciation of social power, and the asymmetries which derive from class, race and gender differences. Thus, the critical notion of 'community' is a conceptual cipher in the utilitarian scheme, which assumes that all social units, however defined, are politically and culturally cohesive bodies which can articulate the sum of their constituents' individual desires and needs as a univocal expression. The aggregate of individual preferences achievable in a market is thus confused with social and political consensus – the market-aggregate expedient with political principle. It seems hardly necessary to point out just how far removed such an assumption is from human reality without turning back the sociological clock to the 1960s when a range of theorists opposed the banalities of consensual social theories such as Parsonian structural–functionalism (see Jary and Jary, 1991).

Bullard (1993c) has described the pressure which waste corporations in North America have placed on Native American communities to accept toxic landfills on their lands in return for compensation. It is not difficult to imagine the allure which financial compensation must hold for indigenous leaders aware of their people's desperate need for economic security. Indeed, Cutter (1995) reports a recent example where the Apache nation sought the establishment of a private nuclear waste facility on its territory in New Mexico in return for monetary compensation. But what power do indigenous leaders wield in the decision-making processes which lead to such results? And what is the potential for other viewpoints to emerge from within such communities where the awareness of industrial risks may be greatest among younger, or otherwise less influential, members? Is it reasonable for indigenous or other minority groups to expose themselves to risk in return for money? What of indigenous ecological rationality, with its much-cherished (by many Greens) insistence on the intrinsic value of nature? This issue exposes the dangers of value relativism that we earlier warned of. Relativism is encouraged by communitarian perspectives that eschew any consideration of how power is distributed within cultural groups.

We share Harvey's (1996) view that the problem of economic insecurity should first be solved for minority groups as a means for avoiding such dilemmas. The seemingly cavalier predilection of some North American indigenous communities for waste facilities must be understood as the product of a decision framework which Heiman (1996: 119) describes as 'the forced fight between jobs and environment'. We say that no minority community should be forced through structural underdevelopment into such invidious decision scenarios where leaders are encouraged to trade their people's environmental health in return for basic material security.

Finally, the utilitarian approach has a much more insidious, systemic consequence in that it depoliticises the production of risk by focusing social attention on distributional issues (i.e. optimal siting patterns for LULUs), which are then handed over to the administrative state. The approach undermines the critical and reflexive potential of the environmental movement, reducing 'fairness' to a question of geographical (or social) equity in allocation of risks. In short, the entire question of industrial risk is reduced to a locational problem, a dilemma over the siting of waste output, which the state must arbitrate as an interest group conflict (Lake and Disch, 1992). In this sense utilitarian solutions to the LULU problem defend what Beck (1995) has termed the 'organised irresponsibility' both of industrial capital, and of the technological bureaucracy that is meant to monitor and control its activities (Heiman, 1996: 120).

The conceptual problems with the utilitarian approach discussed above are beginning to reveal themselves in political practice in North America and Europe whose political landscapes are becoming largely, if not yet universally, frictional for the developers and operators of risk-producing land uses. Clark and Smith have described this dilemma of universal resistance for the waste management industry in the developed world:

> The days are gone when both local publics and governments alike would tolerate waste disposal within their respective boundaries, especially if that waste is being imported from outside the region or locality. The body politic has become more educated as to the dangers inherent in such activities and has ensured that the disposal of waste is no longer a simple matter of finding a suitable hole in the ground.
>
> (Clarke and Smith, 1992: 2)

Vexed industrialists and state technocrats now characterise the ever extensive hostility towards LULUs as 'NIABY' (Not-In-Anyone's-Backyard) (Heiman, 1996), or even 'NOPE' (Not-On-Planet-Earth). What are the consequences of this increasingly pervasive opposition to LULUs for the risk society? First we must ask *which* risk society?' because the growing resistance of local communities towards LULUs has not cohered – at least not yet – as a political movement which can transform the hazardous nature of industrial production itself. Firms are still relatively free to produce, only now they must find new places, outside 'the nation of NIMBYism', in which to do this. As Goldman observes, the increasingly pervasive and potent mood of NIMBYism across the United States has meant that, 'corporations are even more likely to move the most noxious plants to less developed countries, where even poorer communities of color will be the hosts' (1996:

128). In this way, of course, risk-producing firms escape both inhibitive environmental regulation and social resistance to their presence.

What are the consequences of this export of risk? Can there be any global environmental justice when developed countries export their waste and hazardous industries to underdeveloped regions? In the next section we examine these questions, and other ethical implications which arise from the international traffic in risk.

JUSTICE WITHIN THE GLOBAL ENVIRONMENT

States and environmental justice

There is an implicit assumption in most of the debates on justice *in* the environment that the existing political system can in fact deliver it. Largely absent is the question of the justice *of* the political system itself. In fact the global political system, composed of competing nation states, may be unfitted for the task of guaranteeing environmental justice. This question, as we saw in the last chapter, is an increasingly important one in philosophical discussions of justice. It is also highly salient in the debate on justice *to* the environment, justice to nature, which we consider in the next two chapters.

Both trade liberalisation and economic globalisation have allowed firms greater discretion in deciding both where to locate their production activities and in what places to dump the wastes which arise from these. In particular, both the General Agreement on Tariffs and Trade (GATT) and the North American Free Trade Agreement (NAFTA) have undermined the capacity of individual states to regulate trade for environmental ends. This fact was underlined in 1991 when a GATT dispute settlement panel (a 'discursive design' of sorts) used the agreement's 'trade disciplines' to rule that a prohibition on tuna imports by the United States government was inconsistent with the GATT. The prohibition had an explicit environmental objective – to stop the import of tuna caught with purse-seine nets. These devices, sometimes referred to as 'driftnets', tend also to ensnare and kill dolphins at prodigious rates (see Horwitz, 1993; Magraw, 1994). None the less, the regulation was ruled to contradict the GATT's central aim of liberalising world trade. (The action against the U.S. restrictions was brought to the World Trade Organization by the Mexican government; Mexican fishers use driftnets and the state was understandably anxious about the US measure's effect on export income.) As Magraw (1994) points out, GATT does not even contain the word *environment*, and its ruling panels are not required in their deliberations to take account of environmental conventions outside GATT and customary international law. Environmentalists fear, therefore, that GATT and other trade liberalisation agreements, will be used both to dilute environmental regulations in developed countries and to encourage ecologically destructive economic activities in the developing world (Kelsey, 1995; Pulido, 1996). Disputation under GATT has already highlighted the agreement's potential to undermine environmental regulations in developed nations. Note that, to 1994, the United States has been the plaintiff or defendant in every GATT environmental dispute (Charnovitz, 1994).

The competition between states for productive investment within the new globalised economy is, of course, marked by the structural differences between developed and developing countries (and we include many of the former Soviet Bloc states in the latter). We have seen that current measures of investment (by GDP) do not provide a measure of the *value* of that investment. However, deceived by the allure of GDP growth (as currently measured) and impelled by greater socio-economic need, the developing countries have proved far more willing than western nations to accept risk-producing investment, in the form of hazardous industries and imported wastes. The weak and poorly resourced environmental regimes of many developing nations reflect this prioritising of economic security over ecological quality. This structural difference of economic needs and government regulation between the developed and developing worlds, and the absence of any supra-national body to ensure a consistency in environmental standards, has encouraged western industrial capital to shift unpopular and increasingly illegal hazard-producing activities and wastes across national boundaries to states which often define, and welcome, these transfers as 'investment'. We refer to this phenomenon as the 'traffic in risk'.

The traffic in risk

A broad tradition of social scientific analysis has examined the significant growth of investment by western industrial capital in developing countries which has occurred since the Second World War. Many of these analyses have emphasised the relative cost advantages (notably, the cheap supplies of labour power and raw materials) of developing countries for firms eager to escape the perceived disadvantages of developed industrial regions in the west (especially, labour militancy, low productivity and high wages; see, for example, Massey, 1984). Fagan and Webber (1994) point to other motivations behind the shift in western industrial investment patterns, especially after 1970. Their analysis stresses that much of the investment undertaken by multinational capital, transnational corporations (TNCs), outside the developed 'core' states 'was designed to serve the growing domestic markets in [developing countries] rather than for export' (Fagan and Webber, 1994: 37).

As Smith (1984; 1994) and others have argued, this post-war shift is part of an established historical pattern, involving the 'see-sawing' of productive investment between declining and emergent regions and states, which has been a key feature of industrial capitalism since its genesis in the eighteenth century. The importance of this analysis is that it shows that, rather than being a system tending towards equilibrium, as utilitarian economists generally assume, capitalism depends at core on the maintenance of what complexity theorists term a 'far-from-equilibrium' condition or, in Marxist terms, 'uneven development' (Anderson *et al.*, 1988; Brian, 1990; Smith, 1994: 649–50; Fagan and Webber, 1994).

Environmental risk – the threat to human health and ecological well-being posed by industrial capitalism – is a critical dimension of this process of uneven development. Environmental consciousness grew in the nineteenth century, following prolonged industrialisation. A series of movements amongst and between middle and working classes demanded the improvement of sanitary conditions, housing and general amenity in British cities. In the twentieth century, this critical consciousness has deepened and generalised within advanced capitalist nations, finding political expression in the form of increasingly elaborate regimes of state environmental regulation. Thus, ecological consciousness, Beck (1992; 1995) argues, appears as a social contradiction within capitalism (cf. Fagan and Webber, 1994: 35). Yet, as he acknowledges, the (as yet inchoate) age of reflexive modernity is itself developing unevenly, being largely confined thus far to the advanced capitalist nations which have experienced long periods of industrial capitalism. This uneven development of socio-political resistance to risk among nation states further explains the relocation of industrial production from developed to developing regions and countries in the post-war period.

The perceived ecological 'over-development' of the west is intensifying, witnessed in the increasingly pervasive hostility of local communities in advanced capitalist nations towards risk-producing land uses discussed above (see Kemp, 1990; Smith and Blowers, 1992; Johnsen, 1992; McDonell, 1991; Szabo, 1993; Greenpeace New Zealand, 1994). The dilemma of the disposal of the Brent Spar oil rig (discussed in Chapter 1) is a case in point. Given these centrifugal social and regulatory pressures, it should be no surprise that environmental organisations are reporting a flourishing trade in toxic wastes, exported mainly from developed countries to developing nations. Disturbingly, this traffic in risk involves both western waste-producing firms *and western governments*, the latter seeking to dispose of hazardous industrial residues which their own regulations and polities will no longer accept (Smith and Blowers, 1992).

Pulido (1996) has observed that the political successes of the environmental movement in developed countries may actually accelerate the relocation of hazardous industries to developing nations. Environmental regulations are increasingly cited by US firms as a reason for their flight to more 'business friendly' countries, such as Mexico. The profits from industrial plants, as well as their products, are largely exported to the country of the operating firm. The developing nation which hosts the facility retains a quantum of wage and land rent income, but incurs some input expenditure and the risks and consequences which attach to the hazardous industry. A critical aspect of this system of flows, involving the circulation of money, products and risk, is the fact that it permits developed countries to externalise industrial risks by moving hazardous forms of production beyond their borders. In such instances, firms enhance their profits through the imposition of 'cross-border externalities', given that the nations which host hazardous production may never be fully compensated for the spillover effects of these activities (i.e. environmental degradation, social dislocation).

Seen more broadly, the traffic in risk, particularly the transfer of hazards from developed to developing nations, involves more than simply the export of toxic wastes. The entire capitalist commodity system is infused with risk. At all points of production, circulation and consumption there are hazards for human beings which have a variety of sources, ranging from the production and use of dangerous substances, hazardous forms of packaging and distribution and finally, the risks which attach to the consumption of frequently despoiled or faulty commodities.

Within this 'continuum' of risk there are, however, two areas of extreme potential hazard which emerge from separate 'moments' of the production process. First, the process of production itself can pose a risk to workers and surrounding communities, perhaps best exemplified in the examples of a nuclear power station or a chemicals plant. Even a minor technological malfunction or process mishap can have grave consequences both for humans and the environment in their proximity. Second, there are the risks which emerge 'downstream' in the production process, in the form of residuals which may threaten human and environmental wellbeing. These toxic wastes may accumulate at the point of production or, as is generally required in developed countries, they may be removed to some other site for further processing, storage or disposal. This is not to mention the enormous quantities of hazardous wastes which are daily illegally disposed of, frequently dumped in public domains, such as waterways, landfills, sewerage and drainage systems, and in the countryside. The transfer of such toxic residuals from the point of production has thus generated an entire set of hazardous land uses, including waste transport systems, storage warehouses, landfills and incinerators.

The attraction for firms of the developed world is the chance to expand their aggregate output (and thus profit) by operating autonomous plants within national settings characterised by low production costs and large, growing markets. It must be noted that foreign operated plants are often built to higher safety standards than locally owned equivalents, as Hazarika (1987: 22) observes, the worst environmental offenders in developing countries tend to be 'government factories, local private industry and illegal manufacturing units', rather than the plants operated by TNCs. None the less, it is a fact that western TNCs frequently operate industrial facilities in developing countries which are characterised by lower safety standards than those achieved at their own equivalent plants in the developed world. At least some of these conditions were apparent in the production undertaken by Union Carbide at its pesticides factory in Bhopal, India between 1969 and 1984.

Hazardous production and the developing world: the case of Bhopal

In the late 1960s scientists, agricultural businesses and national governments across the globe hailed the arrival of the 'Green Revolution', a radical improvement in crop productivity which would close the Malthusian gap between population growth and food output. The 'Green Revolution' described a radical increase in agricultural productivity that could be achieved through the use of new, high-yielding cereal varieties, fertilisers and pesticides, in concert with specific farming practices (notably, the use of controlled irrigation) (Jones *et al.*, 1990). Not surprisingly, these radical agricultural innovations were greeted with much enthusiasm by states of the developing world. The Indian government was a particular champion of the new agricultural technologies, believing that their use would allow the country to achieve self-sufficiency in cereal production (a goal that was in fact realised). By encouraging the expansion of domestic cereal production, the state hoped to reduce both the incidence of famine and the balance of trade deficit.

For the self-sufficiency goal to be attained it was, of course, critical to produce locally as much of the new agricultural technology as possible. It was with this consideration in mind, therefore, that the Indian government in 1969 allowed – indeed encouraged – the US-based TNC, Union Carbide, to establish a small pesticide production factory in Bhopal, the capital city of Madya Pradesh, one of the country's largest and poorest states (Weir, 1987). The plant was operated by a subsidiary of Union Carbide, Union Carbide India Ltd (UCIL). By the early 1980s, the plant was manufacturing and using highly toxic chemicals, among them methyl isocyanate (MIC), a highly unstable and deadly compound, to produce pesticides such as Sevin and Temik. All the pesticides produced at the UCIL plant were sold in the Indian market.

Shortly past midnight on 3 December 1984, a technical mishap at the UCIL plant caused a large mass, perhaps, as much as 40 tonnes, of MIC to escape from a storage tank into the cool air of the winter night. Soon, a yellowish-white fog began to blanket the sleeping city of 800,000 people (Weir, 1987). The deadly mist quickly settled over the city's crowded slums and squatter colonies, several of which adjoined the UCIL plant (Shrivastava, 1992). Weir recalls the night of terror which followed:

> Hundreds of thousands of residents were rousted from their sleep, coughing and vomiting and wheezing. Their eyes burned and watered: many would soon be at least temporarily blinded. Most of those

fortunate enough to have lived on upper floors or inside well-sealed buildings were spared. The rest, however, opened their doors onto the largest unplanned human exodus of the industrial age. Those able to board a bicycle, moped, bullock, car, bus, or vehicle of any kind did. But for most of the poor, their feet were the only form of transport available. Many dropped along the way, gasping for breath, choking on their own vomit and, finally drowning in their own fluids. Families were separated; whole groups were wiped out at a time. Those strong enough to keep going ran 3, 6, up to 12 miles before they stopped. Most ran until they dropped.

(Weir, 1987: 16 and 17)

By morning it was obvious that the worst industrial accident in history had taken place at Bhopal. A conservative estimate of the human toll one week after the accident included nearly 3,000 dead, 7,000 severely injured, and more than 300,000 others affected by exposure to the deadly mist – some 2,000 animals also perished (Shrivastava, 1992). Pearce and Tombs (1993) put the numbers of permanently disabled at 20,000 (many of these 'disabilities' included horrific disfigurements and painful impairments) while as many as 10,000 people may have died as a direct result of the tragedy. And the suffering continues to the present day. People continue to succumb to maladies which are attributed to the Bhopal disaster. Pearce and Tombs cite one study which has demonstrated that 'of 2700 pregnancies in Bhopal in the year following the disaster, 452 ended in abortion or still-birth, 132 died soon after birth and 30 were malformed' (1993: 192).

Who was to blame for the chemical holocaust at Bhopal? The UCIL plant's safety standards did not match those in Union Carbide's otherwise similar pesticides manufacturing facility in West Virginia. As Weir explains:

Although . . . safety systems are automated with a state-of-the-art computer system at Union Carbide's plant in Institute, West Virginia, which also uses MIC in the production of Sevin and other . . . pesticides, many of the controls at the Bhopal plant were manually operated. Critics charge that this represented a 'double standard,' a characterisation Union Carbide denies. The company says it had specified the design standards for the Bhopal factory, but the actual construction was done by its Indian subsidiary, UCIL, which used local equipment and material. Industry publications say that the Indian government required manual controls wherever possible (1987: 33). Moreover, Union Carbide's chairman at the time of the disaster, Warren Anderson, admitted in March 1985 'that the doomed plant had violated company standards and operated in a way that would not have been tolerated in the United States'.

(Weir, 1987: 59)

Some Union Carbide officials have claimed that sabotage may have been involved in the disaster, though this claim is widely disputed (Pearce and Tombs, 1993). In addition, a combination of sloppy maintenance procedures and several critical design defects had, by 1984, rendered the plant a hazardous time bomb ready to be triggered by malfunction, sabotage or natural disaster (see Banerjee, 1986 on this). In 1982, an investigative team from Union Carbide's US headquarters identified 61 hazards in the plant, 30 of them considered to be major, but the company seems to have ignored this warning (Pearce and Tombs, 1993). Bogard (1989) also highlights the laxity of the safety procedures which Union Carbide had set in place in order to warn the communities surrounding the plant of any accident. For their part, authorities from the state of Madya Pradesh also ignored repeated warnings from an investigative journalist, Raj Kumar Keswani, about the hazardous nature of the UCIL plant.

Regardless of what or who actually triggered the chain of events which culminated in the release of the MIC cloud over Bhopal, Union Carbide, and to some extent, the Government of India, must take responsibility for the tragedy. Bogard lays the blame for the disaster at the feet of the plant's owners and Indian officialdom: 'Union Carbide itself was responsible, the government of India was responsible, a technocratic class that predictably elects profitable, low-cost, high-tech answers for human misery was responsible. Theirs is a responsibility grounded on intentional ignorance, deliberate omission, and misguided optimism' (1989: x). Thus, as Beck (1995) explains, industrial capital, state risk-management bureaucracies, and national legal systems collude in the systematic production of risk landscapes. Moreover, Beck's thesis on the difficulties in attributing responsibility for ecological catastrophes, given that most national legal systems institutionalise 'organised non-liability' for risk, is given support by the compensation settlement reached between Union Carbide and the Indian government.

Originally, the Indian government sought $US3.3 billion from Union Carbide as settlement for the physical damages and human suffering it had caused at Bhopal. As Pearce and Tombs (1993) explain, this figure was anything but excessive when considered alongside the damage claims sought in other recent large-class actions against TNCs and other industrial conglomerates (e.g. the $US2.5 billion received by the 195,000 victims of A.H. Robbins's Dalkon Shield). In February, 1989, the Indian government, acting on behalf of the Bhopal victims, settled out of court with Union Carbide for $US470 million. The problems with existing procedures in international law have already been discussed in relation to the Ok Tedi dispute (Chapter 1). By settling the matter out of court, Union Carbide avoided establishing any damaging legal precedent or liability. It should be noted that Union Carbide was earlier successful in having the case tried in India, rather than in its home country, no doubt fearing the proclivity of the US courts to award large, punitive settlements against corporations which had grossly offended against the public interest. For the US courts, it was by no means straightforward where and how a transnational corporation should be tried for environmental offences. Indeed, as the Indian government argued when seeking to have the Bhopal case heard in a US court, multinational capital is able to use its deterritorialised organisational structure to maximise the advantages of the 'organised non-liability' to which Beck (1995) refers.

There have been many accidents arising from industrial production by western multinationals in developing countries, though none have been as dramatic as the Bhopal tragedy, and few have immediately caused loss of life (though we cannot ignore the real possibility of unverified deaths and injuries). The harmful effects of such plants on developing countries are more often subtle and unseen, as Weir notes:

> Bhopal is being repeated, not just as explosions, infernos, and deadly clouds heard, felt, and seen, the world over, but as 'mini-Bhopals' – smaller industrial accidents that occur with disturbing frequency in chemical plants in both developed and developing countries. Even more numerous and deadly are the 'slow-motion Bhopals' – unseen and chronic poisoning from industrial pollution that causes irreversible pain, suffering, and death.
>
> (Weir, 1987: xi–xii)

Weir documents several of these 'unseen Bhopals', some of which involve western-owned plants producing the same sort of deadly chemicals as at Bhopal and in similarly unsafe circumstances.

The traffic in risk can also involve the relocation from western to developing countries of obsolete industrial plant. In some cases this equipment may have been formerly used in the country of origin to produce chemicals or products which may have been subsequently banned for environmental reasons. Weir reports the case of one Californian chemical manufacturer which shipped its disused DDT formulation equipment to an Indonesian pesticides firm in 1983 (DDT was banned in the United States in 1972). Soon after this, the plant was being used in a village south of Jakarta to produce DDT. By late 1984 locals and environmental activists claimed that pollution from the factory had killed twenty-five villagers and numerous domestic animals (Weir, 1987).

THE WASTE TRADE

Another dimension of the traffic in risk is the toxic waste trade. Indeed, Beck (1995: 134) believes that the 'worldwide traffic in toxic and harmful substances' is a defining characteristic of the present age: the 'risk society'. As he puts it so evocatively, 'Supranational groups of regions and countries swallow poisons and waste on others' behalf' (*ibid.*: 154).

In 1990 the United Nations estimated that the world was producing between 300 and 400 million tonnes of hazardous wastes annually, about 98 per cent of which was generated by OECD countries (Greenpeace International, 1994). A combination of regulation, NIMBY opposition and technical necessity means that much of these wastes must be shifted from where they are generated to other places for storage and/or destruction. An inestimable proportion of the world's toxic wastes are dumped illicitly, often surreptitiously in city drainage systems, the open sea or the countryside.

Some of the trade occurs within the developed world. In the late 1980s, for example, it was estimated that 100,000 waste transfers occurred annually within Europe (Smith and Blowers, 1992). However, a significant amount of the commerce in wastes involves transfers of domestic and industrial refuse (both toxic and non-toxic) from developed nations to poorer countries (*ibid.*: 212). According to Greenpeace, Germany is the largest waste exporter in the world, and in 1993 shipped over 600,000 tonnes of hazardous wastes to ten different countries in Europe (including former Soviet Bloc nations) and to the developing world (Edwards, 1995a). The United States in 1992 exported over 145,000 tonnes of toxic wastes abroad, with large amounts being shipped to Canada and Mexico (Edwards, 1995b). Even relatively minor advanced capitalist nations like Australia and New Zealand exported significant quantities of hazardous waste to Asia (New Zealand's 'toxic colonialism' (*Otago Daily Times*, 10 March 1994: 7), Australia's 'toxic trade' in exhausted lead/acid car batteries (Daly, 1996)). Smith and Blowers (1992) detail the export of wastes, some of which included radioactive materials, by both the United States and European countries to Africa during the late 1980s with Guinea-Bissau as a major destination – that country being offered the equivalent of its then existing GNP (some $US120 million) 'to dispose of European hazardous waste in landfills' (*ibid.*: 212). Smith and Blowers also report the growth of waste trading *between* developing countries.

Under pressure from environmental lobby groups (notably Greenpeace), European governments agreed during the late 1980s and early 1990s greatly to restrict further exports of waste to the developing world. Developing nations have also imposed controls. The 1991 Bamako Convention, for example, achieved an Africa-wide ban on waste imports (Greenpeace International, 1994). By the early 1990s, a similar ban was in place covering Central America (Greenpeace International, 1996). The principal international waste trade regulation is the 1989 Basel Convention on the Control of Transboundary Movements of Hazardous Wastes.

However, Smith and Blowers (1992) point out that these international controls have thus far proved inadequate given the huge economic incentives which continue to stimulate the traffic in waste. Both governments and private firms can profit enormously through the waste trade. Not surprisingly, these authors conclude that 'a network of waste "brokers" is already operating in Europe and these economic incentives will become prime factors in encouraging such entrepreneurs towards profit maximisation by seeking a Third World location for waste' (Smith and Blowers, 1992: 221). Waste entrepreneurs have been able to exploit a critical loophole in the Basel Convention which permits exports to developing countries if the waste is destined for 'recycling'. Lax trade regulations have thus allowed many waste-exporting firms to ship toxic materials to developing countries disguised as 'garbage' or waste for recycling (Greenpeace International, 1994). In 1996, for example, China accused the United States of violating the Basel Convention by illegally exporting 'garbage' containing radioactive waste to Chinese dumps (*The Canberra Times*, 3 July 1996: 9).

In 1994 parties to the Basel Convention agreed to the immediate ban of all hazardous wastes exports from OECD states to non-OECD countries (Greenpeace International, 1996). This measure sought to close the 'garbage for recycling' loophole by banning the shipping of all wastes for recycling to developing countries after 31 December 1997. Greenpeace hailed the move as a 'victory for environment and justice', but warned that 'there are still a few governments together with cohorts from industry who are still intent on undermining the ban decision' (Greenpeace International, 1996: 1–2). Indeed, Edwards (1995a: 13) reports that Germany, Australia and Britain are trying to undermine the ban by signing bilateral agreements for the trade of recyclable wastes with developing nations. In any case, many of those involved in the waste trade doubt the efficacy of the controls established by the Basel agreement. One leading German industrialist has described the convention as 'poorly defined and open to interpretation from end to end' (Edwards, 1995a: 13). There have also been a number of episodes in the early 1990s involving ships carrying unwanted cargoes of toxic wastes drifting from port to port in an unsuccessful search for a country willing to take their dangerous loads. Smith and Blowers (1992) intimate that at least some of these instances have ended in the illegal dumping of wastes at sea.

The traffic in waste both undermines political attempts to change the nature of hazardous production in the developed world (hence Greenpeace's policy of opposing the trade) and also exacerbates international inequality by further eroding the wellbeing of developing countries. Altvater (1993) explains the waste trade within a broader geopolitical framework, in which developing countries function as energy mines and entropy sinks; viz., they supply raw materials for western industry and act as waste dumps for the corrupted energy which this production generates. Smith and Blowers (1992) support this characterisation, and point to the role of multinational corporations – including both manufacturing and waste management firms – in a global process of *cascading exploitation* whereby wastes are transferred from core (advanced capitalist) states to both semi-core nations (e.g. former Soviet Bloc nations) and, most especially, the underdeveloped periphery.

As with some poorer communities within nation states, so, internationally, risk-producing capital often uses its economic power over impoverished nation states in a form of environmental blackmail: 'The ability of TNCs [transnational corporations] to circumvent legislation by moving their operations to another country allows them to exploit those countries which are desperate for foreign capital' (Smith and Blowers, 1992: 217). Moreover, so Beck (1995: 154) claims, in the international context, 'Suing for damages helps just as little as protesting publicly'. Once a state enters into the waste trade, there is little chance of turning back: 'Regions swallow not only the poison but also its non-attributability. . . . For on top of everything else, the "poison-swallowing regions" are under compulsion to hush it up'. In the era of the risk society, waste, with its rising exchange value, and waste facilities, with their capacity to generate big profits, now appear as sources of economic 'development' which poorer states compete for within secondary or marginal investment circuits.

Here, then, many of the national environmental injustices which were earlier highlighted are mirrored at the international scale. Outwardly at least, political leaders accept heightened environmental risk for their communities in exchange for financial 'compensation'. Yet the internal distribution of this 'compensation' is also an ethical issue. The ubiquity of authoritarian regimes in developing countries (Burma is an extreme example, but there is a continuum of post-colonial, post-communist authoritarianism) means that national communities are endangered by politically deceitful arrangements between states and waste capital that are designed to further the power and material interests of political elites. In such conditions, there is very little chance that the waste trade can improve even the short-term material wellbeing of the masses. This distinguishes the problem from the trading in LULUs in the developed world where it may reasonably be argued that, in certain circumstances, such exchanges might materially benefit local communities in the short term. In the long term, the wellbeing of *all*, including elites, is *risked* by waste investment.

CONCLUSION

In the normal process of production, accumulation and exchange in the world capitalist system, massive environmental injustice is occurring by almost any criterion except perhaps one: the entitlement to property. Yet that is the standard which still far outweighs all the rest. Let us not be deceived by the publicity given to environmental successes. These are the rare exceptions to an extremely dismal norm of constant, largely unseen, daily degradation of the world's environment.

What can be learned from environmental justice struggles within developed nations? Certainly one thing is the importance of the administrative state to regulate the outcomes of processes and structures that distribute environmental quality. How far a wide range of values can intrude upon the instrumental rationalities of the state is always in question. But the administrative state remains the focus of political struggle in which the dialectic of justice takes shape. However inadequate it may be, the administrative state is the only instrument open to lifeworld values, which is today capable of balancing the power of multinational capital organised by competitive markets. Some, at least, of the seeming inability of nation states to respond to the demands of their constituents, as well as some of the shaping of these very demands, can be laid at the door of the unregulated, competive, globalised, market system which blindly directs 'investment' for short-term profits.

The US 'Environmental Justice' framework may provide a prototypical model for other nations. However, as the foregoing analysis noted, there is an urgent need for international environmental and ecological justice movements to transcend the 'politics of place' in order that the nature of industrial commodity production may itself become problematised. In short, the political critique within developed countries must be shifted from the *spatial allocation* of risk to the *production* of risk. Failing this, the environmental justice movement of the United States, for example, will find itself trapped in the politics of distributional justice (i.e. the LULU problem) that ultimately cannot secure universal justice for all human communities and the global environment. At present, the diverse US environmental justice movement, in concert with a more popular opposition to LULUs, has managed only to 'half transcend' the distributional politics of place. The combined effect of these popular and institutional forces may have created a 'landscape of resistance' for risk-producing and risk-managing industries, but this achievement may only have served to ensure that environmental hazards are exported to more 'accepting' landscapes, including developing countries.

As we have shown, the traffic in waste and other environmentally injurious development is worsening international inequity and helping to sustain risky industry throughout the globe. The absence of a supervising state – and the United Nations in its present manifestation cannot yet perform this role – means that a distributional framework cannot be readily applied to the international traffic in risk. Those international agreements which have sought to control aspects of the traffic in risk, such as the Basel Convention, have been shown both to be vulnerable to political attacks by recalcitrant states and difficult to enforce. Epochal structural changes, including economic globalisation, the mobility of capital (and risk) and the collapse of Cold War antagonisms, have created a new geo-political context for ecological politics:

> We are on the threshold of a new phase of risk-society politics; in the
> context of disarmament and the relaxation of the East–West tension,
> the apprehension and practice of politics can no longer be national but

must be international, because the social mechanism of hazard situations flouts the nation-state and its systems of alliance.

(Beck, 1995: 162)

Indeed, we argue that this new international political practice, of which Beck speaks, must seek to eliminate the flourishing traffic in risk which is already worsening the legacy of global uneven development bequeathed by centuries of colonialism and capitalism. This new ecological politics requires a new global institutional context which can both problematise the production of risk and regulate the distribution of hazards between states. We consider the problematics of such an institutional context in Chapter 7.

More immediately, a question we have not thus far addressed concerns an aspect of 'the environment' we have had to take for granted: its quality. An environment, 'the' environment has value. The question is where this value comes from. If we are to go beyond 'environmental justice' to 'ecological justice' (justice between humans and non-human nature) we will have to consider this question, to which we now turn.

6

ECOLOGICAL JUSTICE
Rethinking the bases

The maxim 'live and let live' suggests a class-free society in the entire ecosphere, a democracy in which we can speak about justice, not only with regard to human beings, but also for animals, plants and landscapes.

(Naess, 1989: 173)

INTRODUCTION

The distribution of environmental quality is the core of 'environmental justice' – with the emphasis on *distribution*. The instrumental interest people share in having a safe, healthy and pleasant environment in which to live is unproblematic. Ecological justice is a diffferent matter. Here we have to consider the meaning of environment in a deeper sense, the sense of our moral relationship with the non-human world. Of course, the two senses are related in that the value of the environment is changed and considerably extended if the relationship is construed not just as an instrumental one but a moral one.

In modern philosophy justice has for the most part been conceived in terms of the relationship of self to other humans – though Kant, Bentham and Marx, whose ideas we have already discussed, and Spinoza, who we discuss in this chapter, all considered human–nature relations. The passage from Seneca in the front of this book shows that the idea of justice to nature is an ancient one. To conceive of justice to nature, ecological justice, it is necessary to reconceive of the basis of justice in the way we think of our 'self' and thus how we define our interests and moral values. As we have seen, a variety of different ways of conceiving of justice has emerged as different strands of thought and social movements have placed different challenges to our picture of 'the self' on the political agenda. This picture has been reshaped and reinterpreted several times in the last two or three hundred years. Perhaps the most profound change in this century has been the challenge to individualism in which greater account is taken of the person as a member of a culture and a society. But individualism, reacting against feudal societies, and latterly communist societies, itself supplied a revolutionary redefinition of the person as transcending community and society. In the most extreme version of this picture both community and society are made to vanish entirely. Culture becomes an aggregated and homogenised sphere of individual tastes. This dialectical process of shaping and reshaping our idea of the self has not ceased. And the direction we think it is already taking today is one in which the horizons of the self (to use Charles Taylor's powerful phrase) are widening once again to embrace both community and environment. This is the view we find in modern ecophilosophy.

As with any major political movement there are differences of view within ecophilosophy. Different strands of thought emphasise different aspects of the ecological project. A pivotal question throughout, however, is whether 'justice to nature' means abandoning the principles of western liberal philosophy or merely their modification. A definition of self which ignores the need for individual difference and differentiation, stressing only identification with community and environment, is, we think, as diminishing as a definition which ignores the need for community and environment and reduces the self to an isolated pinpoint. Is an ecologically enlarged conception of the self compatible with the discriminations we customarily make for purposes of justice? Does expanding the self-picture mean abandoning conceptions of justice which take human persons as ends-in-themselves? In this chapter we first consider the idea of the self. We then discuss three ways in which ecophilosophers argue for an expansion of the self's horizons. Finally we suggest ways in which the bases of justice might be reconstrued in the light of these expansions.

EXTENDING THE SELF-PICTURE

Conceptions of justice deal with relationships: how 'I' am to be morally related to the world around. If the self is conceived as a bounded 'ego' isolated from the world around, then the good of the self, self-interest, is entirely constituted from within, subjectively. Whatever is outside the self can act as an input, can 'do good' or 'do bad' to the self (or for that matter 'be' good or bad). But whatever happens to anything outside the self can have no *direct* impact on the *self's* welfare. It may of course have an indirect effect if the event changes the inputs. This picture of the self helps make sense of Kant's view that only the performance of 'duty' can be regarded as moral conduct. Duty is precisely those acts which do not benefit the bounded self. No wonder, then, that the moral imperative had to be categorical, unconditional. Kant did not challenge the picture of the 'bounded self' and therefore all morality appears as a burden. Justice in these circumstances is about the fairness of transactions between bounded selves when no 'return' can be expected from a particular transaction.

The bounded self has the advantage of simplicity and calculability. It is the self of the 'self-interested' individual of Thomas Hobbes, Immanuel

Kant, Adam Smith and Jeremy Bentham and it is the self-picture behind mainstream utilitarian economics. It is the self-picture of hedonism (Bell, 1976), narcissism (Lasch, 1978), moral subjectivism and, to some extent, the participants in the social contract of Rawls (1971). The market is supposed by neo-liberals like Hayek (1979) to dispense with the need for a public morality of mutual care. If values are constituted only subjectively from within, then Hayek is correct to argue that ultimately social justice has no meaning. If we live only for the moment, knowing neither past nor future, knowing only our internal material needs of the moment, then greed is simply a legitimate expression of the self. This picture of the self tends to foreclose the discussion of justice. If all that we are is constituted from within, then the best that we can hope for is indeed a mechanism such as the market to negotiate self-interest without resorting to war. A system of production based entirely upon greed is justified.

However, this picture is changing. The change is particularly clearly illustrated in the thinking of a leading philosopher of the 'New Right'. In 1974, in *Anarchy, State and Utopia*, Robert Nozick took the idea of the utilitarian bounded self to its logical conclusion in a privatised market-organised society. In *Philosophical Explanations* (1981) he problematised this self-picture but, in thinking about morality, did not go much beyond a Kantian 'subsumption' of the bounded self. In *The Examined Life* (1989: 258) he takes a critical step further: 'To be related to the deepest reality, in the sense we have described', he says, 'is to embody and exhibit it, something one can do through one's *own* characteristics' (author's emphasis). Now problematised is 'the deepest reality'. He explains what he means:

> I think what is important is to offer responses as something *due*, to respond to things as homages to their reality. What would matter, then, would not be the quantity of our response, even the quantity (or bulk) of the response's reality, but the manner of the response, the spirit in which it is done. Speaking of what is 'due' may make it seem like a debt owed, though, or an obligation, whereas I mean something more like applause. Or an offering. Or, perhaps, more like love. To love the world and to live within it in the mode this involves gives the world our fullest response in a spirit that joins it. The fullness of this response enlarges us too; people encompass what they love – *it becomes part of them as its well-being becomes partly theirs*. The size of a soul, the magnitude of a person, is measured in part by the extent of what that person can appreciate and love.
>
> (Nozik, 1989: 258, emphasis added)

The bounded self is the self-picture with which we enter into the world as a baby. It is not the self-picture of a mature human adult who has experienced both the flux of time and relationships with other people. Charles Taylor (1991) has explored what he calls the 'inarticulate debate' about the nature of 'authenticity', that is the debate about the constitution of the self. His point is that if the pursuit of authenticity is nothing but seeking more and better inputs to the bounded self, then critics of modernity like Bell and Lasch are right to condemn the pursuit of authenticity, being true to ourselves, as hedonism and narcissism. But, as Taylor puts it:

> Just because we no longer believe in the Aristotelian doctrines of The Great Chain of Being, we don't need to see ourselves as set in a universe that we can consider simply as a source of raw materials for our projects. We may still need to see ourselves as part of a larger order that can make claims on us.
>
> (Taylor, 1991: 89)

In fact Taylor leads a body of evidence from poetry and philosophy to show that we do in fact see our 'selves' in this way. If this is so, 'If authenticity is being true to ourselves, is recovering our own *sentiment de l'existence*, then perhaps we can only recover it integrally if we recognise that this sentiment connects us to a wider whole' (*ibid.*: 91). 'Indeed', he writes, 'It would greatly help to stave off ecological disaster if we could recover a sense of the demand that our natural surroundings and wilderness make on us' (*ibid.*: 90).

Within the ecological movement the debate is far from inarticulate. By the time the first pictures of Earth from deep space were flashed around the world and placed in front of a large proportion of the human race, consciousness of the fragility of the planetary environment was already awakening (following, for example, Carson, 1962). These images helped make concrete the wider whole to which the person could relate. Roszak explicitly connected the pursuit of personal authenticity with the welfare of the planet. 'What then does the Earth do?' he asks:

> She begins to speak to something in us – an ideal of life, a sense of identity – that has until now been harbored within only an eccentric and marginal few. She digs deep into our unexplored nature to draw forth a passion for self-knowledge and personal recognition that has lain slumbering in us like an unfertilized seed.
>
> (Roszak, 1978)

Roszak's argument is that self-fulfillment is now a mass pursuit, where formerly it was limited to a small privileged minority, and that self-fulfillment involves identification both with the small 'situated' human group and with the larger planetary whole.

There are, however, different views and vigorous debate about the constitution of the self. In his examination of ecological ethics Fox (1990) points to a division between those philosophers who seek to expand the scope of the moral environment and those who seek to expand the scope of the self. He is wrong however to reduce the debate to psychology. The debate is about what we *are*, not about the nature of an individual psyche: ontology not psychology. In fact we think that ecophilosophy has a common underlying theme which is an expansion of the self-picture; a reaction against the picture of the bounded self which governs today's dominant economic institutions. Ecophilosophy points out that the bounded self has become so tightly restricted and internalised that it is a distortion even of the philosophy of the Enlightenment in which the picture originated. They wish to reinstate what they see as a truer picture of the self connected with its environment. Ecophilosophers find different ways of understanding this reinstatement. We will here discuss three such ways: the expansion of the moral environment, the expansion of the social environment, and the expansion of the self.

EXPANDING THE MORAL ENVIRONMENT

Kant postulated that human beings must not be regarded as instrumental means for any purpose whatsoever. As we have seen, this is a central axiom of most modern theories of justice. It has served well as the basis of a humane and compassionate ethic. Intrinsic value theorists seek to extend the 'kingdom of ends' to the non-human world and thus to break down the moral barrier between humans and non-human nature. This step is problematic for two reasons. First, we are accustomed to thinking of humans as the primary source of value. 'Values', in our modern humanist world, come from *human* conduct. If the source of values is somewhere else, then where is it? Do we have to return to an acceptance that values are given to humans by some supra-human source, be it through mystical communion or through scriptural revelation – as, for example, in forms of Buddhism, Taoism or the Christianity of Teilhard de Chardin? Second, if we can find an acceptable non-anthropocentric source of values, what consequences will this have for our conception of justice?

There is an odd behavioural inconsistency in western liberal societies. On the one hand we regard civilised behaviour as necessitating a moral attitude of *respect* towards animals and *care* for nature. Such an attitude has a long history. Yet, on the other hand, we condone the utterly inhumane treatment of animals *en masse* and the devastation of ecosystems when it happens out of sight and out of mind. This inconsistency has been noted not only by ecologists (Naess, 1989; Midgley, 1992) but by the psychologist Ronald Laing. The 'objective look', he says, can encompass unimaginable horrors without flinching, in the name of science. The absolute divorce of facts and values, and science from experience feeds this inconsistency:

> Having decided that the knowledge of good and evil is not what it (science) knows, or aspires to know, it will tell us what to do, glad to be ignorant of spirit, mind and soul, love and hate, beauty and ugliness, and everything that most people suppose makes life worth living.
> (Laing, 1982: 23)

A first point to note, then, is that *maintaining* the radical moral distinction between humans and animals is itself looking a little dated. Intrinsic value theorists, it might be said, want ethics to absorb the insight of Darwin's theory of evolution. If the origin of species is a process of evolution (however it works in detail) and if all species, including humans, have a common ancestry, then we can no longer use the species 'human' to demarcate an absolute ethical frontier. Regan (1983) argues that the Kantian intuition of 'inherent value' should be defined not in terms of some necessarily arbitrary species demarcation but in terms of who is 'the subject of a life':

> Individuals are subjects-of-a-life if they are able to perceive and remember; if they have beliefs, desires, and preferences; if they are able to act intentionally in pursuit of their desires or goals; if they are sentient and have an emotional life; if they have a sense of the future, including a sense of their own future; if they have a psychophysical identity over time; and if they have an individual experiential welfare that is logically independent of their utility for, and the interests of, others.
>
> (*ibid.*: 264)

Being responsible for one's actions, that is being a moral 'agent', Regan argues, has never been the primary qualification for moral considerability. Subjects-of-a-life can be passive subjects, or patients. There are many human examples of moral patients. In fact all humans go through a stage in their lives of moral-patienthood (childhood) from which we gradually emerge as 'agents' – and to which we sometimes return in old age. Those animals which can be considered subjects-of-a-life must be treated as moral patients. The consequence is that although we may, as humans, defend ourselves against animals we should not, for example, eat them to satisfy our tastes or use animals as live instruments in laboratory experiments. Singer (1975; 1979; 1985), employing the simpler utilitarian criterion of whether an individual can suffer pain, takes a similar position.

So Regan and Singer have moved the moral frontier to include our nearer animal relatives. They have not abolished it. There is still the frontier between animals which we can identify as the 'subject-of-a-life' and other creatures. Regenstein (1985) argues that criteria such as Regan's are too restrictive. Animals with which we have difficulty identifying (snakes, alligators, for example) are also morally considerable.

The ideas of Aldo Leopold have been extremely influential over a broad spectrum of the environmental movement in North America. (for example, 'Earth First', the American version of deep ecology, the global integrity movement; see Pepper, 1996: 25, 51). Leopold's (1949) 'land ethic' extends moral considerability yet further – beyond animals to biotic communities. 'We can be ethical only in relation to something we can see, feel, understand, love, or otherwise have faith in' (*ibid.*: 214). He takes a communitarian position, thus:

'Ethics are a kind of community instinct in the making. . . . The land ethic simply enlarges the boundaries of the community to include soils, waters, plants, and animals, or collectively the land' (*ibid.*: 204). 'A land ethic changes the role of Homo Sapiens from conquerer of the land-community to plain member and citizen of it. It implies respect for his fellow members, and also respect for the community as such' (*ibid.*). An ecological conscience, he says, is an internal conviction of individual responsibility for the health of the land. His classic statement of the land ethic is simple: 'A thing is right when it tends to preserve the integrity, stability, and beauty of the biotic community' (*ibid.*: 224).

This is not really a departure from anthropocentrism: simply humans experiencing a *feeling* of responsibility, a *feeling* of citizenship or partnership with the rest of nature. But this is a weak account of value, one which Kant wanted to avoid because it depended on subjective feelings which would vary among humans. Such a view would not confer rights on the non-human world. Non-human nature would depend on something akin to human charity. The ecologist J. Baird Callicott (1985), accepting the subjectivist ethics of G.E. Moore (1903), holds that 'value' can *only* arise from the internal experience of the 'valuer' (the atomistic, bounded self).

Pragmatists also find the source of value in humans but the focus is in conduct rather than subjective experience (see Light and Katz, 1996). Saying that something has value is no more than a linguistic expression. What comes first is the act. So when Kant commands us to value human beings as ends in themselves, what he is really doing is recommending a special form of conduct towards human beings. We behave towards instruments differ- ently from the way we behave towards entities that are 'ends-in-themselves'. The idea 'end-in-itself' is a way of explicating the behavioural difference. That form of conduct can be extended to the non-human world without worrying about what 'value' *is* , which is an unsolvable puzzle even when applied to the human sphere ((Neale (1982) and Weston (1985) discuss this problem.) Fox (1990) takes a pragmatic position in defense of ecocentrism. The argument against ecocentrism that humans can only think like humans is in Fox's view 'weak, trivial and tautological'. Anthropocentrism, 'in the sense that really matters' is 'exhibiting unwarranted differential *treatment* of other beings on the basis of the fact that they are not human' (Fox, 1990: 21). It is thus our *conduct towards* and *treatment* of the non-human world that is the subject of critique.

The question of ecological justice arises from our treatment of the non- human world which is in turn derived from a view about how we are connected with it. Some ecophilosophers have seemed to demand no discrimination among species, indeed at an extreme, a 'biospherical egalitar- ianism'. In this view the whole of nature has an equal right with humans. But this position makes the whole idea of rights self-contradictory. While the 'kingdom of ends' was limited to a single species, humanity, a simple, moral egalitarianism is practicable. Even though actual, material inequality comes to be justified in many different ways, the same moral principles apply equally to all members. This cannot be so if the kingdom is extended even as far as other predatory mammals because these animals live off other animals. Predation is as natural as co-operation. Some species treat other species as instruments of their own survival. In nature itself, therefore, discriminations are made. Humans as animals also eat other species for survival: 'the process of living entails some form of killing, exploitation and suppression' (Fox, 1984: 198).This contradiction is discussed by Plumwood (1993: 172). No one has yet suggested that predators should be prevented from such natural behaviour, so simple moral egalitarianism must be abandoned.

Some ecophilosophers, including Naess (1984), have wanted to ignore the problem by saying that valuing everything equally does not mean we have to treat everything equally. But this will not do. To mean anything at all, ethical *value* (or valuing) must be consistent with our conduct towards others. So conduct which makes distinctions requires moral justification. How, then, can we justify the moral discriminations we do, and must, make?

Benton (1993) acknowledges the necessity of extending the 'ethical circle' to include non-human nature. He argues for a naturalistic Marxism, which draws *inter alia* upon Doyal and Gough's (1991) work on human needs (see above: Chapter 3). However, if the species barrier is broken, he argues, we have to be much more discriminating about the ethical dimensions of the non-human. It is a projection of human tastes (anthropomorphism) to include all those non-humans who are somewhat like humans and exclude all the rest. But Benton reminds us that we are not only *like* animals, we *are* animals. Human powers, Benton agrees, distinguish us from other species, but we share many of their needs, for example the need for health, security, nutrition and shelter. Animals and humans are materially interdependent. Therefore we humans must extend the protections of justice and care to them. Like Turner (1993), Benton sees the universal fact of the frailty of embodiment as a critical source of interdependency for humanity. In extending the idea of interdependence to all creatures, he proposes a new political ontology of a 'human–animal continuum'. But his work does not enable us to make meaningful discriminations within the continuum, and such discriminations will be necessary if we are to extend ecocentric thought into the domain of justice.

Breaking down the moral barrier between humans and animals has its dangers. One is that by so doing we may be tempted to modify our moral standards downwards; to treat other humans more like animals rather than animals more like humans. Both Ferry and Regan, though for different reasons, have warned of the danger of 'ecofascism'. Ferry ([1992] 1995) notes the moral inconsistency between the advanced ecological laws relating to the treatment of animals promulgated by the German Nazi Party, and the behaviour of that party towards humans. Regan (1983: 361–2) condemns the collectivist view of nature and humanity which he finds in the work of Aldo Leopold: 'Environmental fascism and the rights view are like oil and water: they don't mix.' These warnings should not be dismissed lightly. Nevertheless what Midgley (1992) calls '*exclusive* humanism' appears untenable. Once we acknowledge that humans are not the only beings worthy of moral consideration, we can bring what we intuitively know and feel into equilibrium with our reasoning. We can begin to address in moral terms the extraordinary inconsistencies in our behaviour towards animals.

For these theorists the picture of the self is more or less what we have inherited from Kant: a moderately bounded self connected to a moral environment which includes other selves. These theorists seek to expand that moral environment to include aspects of non-human nature. They come from both liberal and socialist positions. Regan, the liberal, would certainly not agree with Benton's collectivist socialism, though the Benton clearly acknowledges the moral importance of the individual sphere. In the next section we consider those whose principal target of critique is the social environment.

EXPANDING THE SOCIAL ENVIRONMENT

Social ecologists seek to expand the scope of society to include the non-human natural matrix in which it is embedded. We have already seen in Chapter 2 that Marx viewed the human self as organically related to the non-human natural world. But he also regarded the idea of 'nature' as a human product. The nature that preceded human history, he says in *The German Ideology*, no longer exists. So the rise of humanity is as much a natural phenomenon as the rise of the dinosaurs. Humans are not external to nature, and 'Nature' is not some pristine state to be preserved from human contact. Indeed to preserve 'wilderness', however desirable that might be, is simply to create another category for the purposes of human intervention.

Engels has sometimes been accused of a Promethean view of the human domination of nature. But he was not insensitive to the results of this domination. He says, 'The animal merely uses its environment; man by his changes makes it serve his ends, masters it' (Engels, [1876] 1995: 74–5). He goes on, however, to observe that each victory over nature takes its revenge on humans. The people of Asia Minor who destroyed the forests to obtain cultivable land:

> Never dreamed that by removing, along with the forests, the collecting centres and reservoirs of moisture they were laying the basis for the present forlorn state of those countries. . . . At every step we are reminded that we by no means rule over nature like a conquerer over a foreign people, like someone standing outside nature – but that we, with blood and brain, belong to nature, and exist in its midst, and that

all our mastery of it consists in the fact that we have the advantage over all other creatures of being able to learn its laws and apply them correctly.

(*ibid.*)

The more we learn about nature the more will humans, 'not only feel but also know their oneness with nature, and the more impossible will become the senseless and unnatural idea of a contrast between mind and matter, man and nature, soul and body' (*ibid.*). This is surely a remarkably prescient view of how knowledge of our relationship with the non-human world has developed.

In the ecosocialist perspective, the individual person, human society and the non-human world are in a threefold process of mutual transformation. Thus, writes Pepper:

> through learning how to farm nature's products, we changed ourselves from nomadic hunter/gatherers to sedentary people. Through learning how to manufacture things we changed ourselves to an industrial society. . . . Through changing nature and making things, we have changed ourselves into creatures who can appreciate the beauty of what we create; buildings, machines, art.

(Pepper, 1993: 112)

So, as the relationship between society and nature further develops, we will see a further change in nature, in society, and in the human self. As we develop the means to learn about nature's vulnerability we may also change our self-perception to one of mutual dependency. But this mutual dependency has to work three ways. Human persons are dependent on other humans, human society is dependent on non-human nature, non-human nature is dependent on human persons and society. The ecosocialist self is therefore one in which the reality of the person–society–nature relationship is fully realised and in which reifications and fetishes such as the 'individual' and 'nature' are done away with (*ibid.*: 126).

Ecosocialism is anthropocentric, the moral relationship between human society and nature is one of 'stewardship' (Attfield, 1983). According to Pepper, 'there is no possibility of a "socialist biocentrism" . . . since socialism by definition starts from concern over the plight of humans' (1993: 224). Ecosocialists thus reject intrinsic value theory for its implicit anthropomorphism (cf. Hayward, 1994) – the charge that biocentrism and deep ecology simply project a (certain) humanly conceived value (intrinsic worth) on to the environment, thus subordinating other critical social priorities, such as justice and self-realisation (see also Grundmann, 1991). However, ecosocialism is not simply anthropocentric in a Promethean way. Ecosocialism's regard for non-human nature reflects the 'enlightened self-interest' advocated by Hayward (1994), stressing the mutuality of human and ecological wellbeing, while recognising a hierarchy of moral significance. Human rights both precede and establish a certain moral significance for nature. Hence Benton (1993) agrees with Regan and Singer that non-human beings are 'moral patients', and should be the objects of human moral concern. Ecosocialists prioritise environmental justice over ecological justice only to the extent of insisting that human social wellbeing is an essential precondition of ecological wellbeing.

Murray Bookchin, the ecoanarchist, conceives of the self as *formed* in action and by virtue of being free to act. Direct action is therefore 'a moral principle, an ideal, indeed a sensibility' (Bookchin, 1980: 47). Bookchin appeals to the Hellenic idea of the citizen as a member of 'the fraternity of selves that composed the polis'. To be a 'self' is to be an individual: competent, intelligent and endowed with moral probity and social commitment (*ibid.*: 9). The moral self is one with 'the capacity to exercise control over social life'. It is the self with personal fortitude and moral probity, without which 'selfhood dissolves into mere egohood, that hollow, often neurotic shell of human personality that lies strewn amidst the wastes of bourgeois society like the debris of its industrial operations' (*ibid.*: 120). Thus for Bookchin, the self is not the bounded self of 'egohood'. Problematic human relationships with nature today do not lie within the person, or even with the person's relationship with nature, but rather with the social systems, which are systems of domination, which shape personal conduct. It is not that people have the wrong ideology, the wrong attitude, but that their activities are embedded in social structures which permit or encourage environmental damage: 'Lumberjacks who are employed to clear-cut a magnificent forest normally have no "hatred" of trees' (*ibid.*: 24). 'To understand present day problems, ecological as well as economic or political' says Bookchin, 'we must examine their social causes and remedy them through social methods.'

Human society is part of the totality of the natural world: 'Social life does not necessarily face nature as a combatant in an unrelenting war. The emergence of society is a natural fact that has its origins in the biology of human socialization' (Bookchin, 1990: 26). However, a feature of this human society is its ability to create a 'second nature' as a cultural tradition and an artificial environment for itself. 'Social ecology tries to show how nature slowly phases into society without ignoring the differences between society and nature on the one hand, as well as the extent to which they merge with one another on the other' (*ibid.*: 30. 'The divisions between society and nature have their deepest roots in divisions within the social realm, namely deep-seated conflicts between human and human that are often obscured by our broad use of the word "humanity" ' (*ibid.*: 32).

Bookchin provides a salutary critique of the antihumanist, misanthropic and neo-facist tendencies of ecophilosophy. He rounds on Naess and his followers. Naess, he thinks, is unoriginal, academic and out of touch with 'groups actively trying to expand public consciousness of environmental hazards' (Bookchin, 1995b: 88). The philosophy of deep ecology feeds the irrationalism and social quietism of the 'Mystical Zone' in United States society and culture (*ibid*.: 92). He points to Foreman's infamous remark that aid for starving people in Ethiopia should be witheld and that 'the best thing would be to just let nature seek its own balance, to let the people there just starve there' (Bookchin, 1995b: 107, citing Devall, 1986). He reminds us of Devall's view that the rattlesnake under the child's bed should be allowed to stay, on the grounds that it has equal rights with the child (Devall, 1990). He cites the implicit racism of Edward Abbey who called for a halt to immigration to the USA on the grounds that 'Latins' would introduce crime and violence as 'normal instruments of social change' (Abbey, 1986; 1988); and the implicit exclusion (by Devall) from the consideration of 'deep' ecology of urban social issues such as gross socio-economic polarisation (environmental injustice).

Biospherical egalitarianism, Bookchin argues, is logically impossible. Human thought and conduct cannot be reduced to the level of, for example, the instinctual navigational behaviour of birds. When we admire the 'skills' of animals we are imposing our own human concept of a skill. This is, he says, anthropomorphism, the projecting on to animals of concepts which are exclusively human: 'Placing human intellectual foresight, logical processes, and innovations on a par with tropistic reactions to external stimuli is to create a stupendous intellectual muddle' (Bookchin, 1995b: 101). The key fact of human, as opposed to animal, existence is that humans have the capacity for knowledge not only of the world around them, the 'environment' in its fullest social and material sense, but also of themselves and their place in this environment. Thus:

> In fact, the ontological divide between the non-human and the human is *very* real. Human beings, to be sure, are primates, mammals, and vertebrates. They cannot, as yet, get out of their animal skins. As products of organic evolution, they are subject to the natural vicissitudes that bring enjoyment, pain, and death to complex life forms generally. But it is a crucial fact that they alone *know* – indeed, *can* know – that there is a phenomenon called evolution; they *alone* know that death is a reality; they *alone* can even formulate such notions as self-realization, biocentric equality, and a self-in-Self; they alone can generalize about their existence – past, present and future – and produce complex technologies, create cities, communicate in a complex syllabic form.
>
> (*ibid*.)

The entire project of ecology, 'deep' or otherwise could not have come about without the special human characteristic of self-knowledge. The salvation of the planet from human excess depends utterly and exclusively upon the characteristic of thought which separates humans so decisively from their closest animal relations:

if we were nothing but 'plain.citizens' in the ecosphere, we should be as furiously anthropo-centric in our behaviour, just as a bear is Urso-centric or a wolf is Cano-centric. That is to say, as plain citizens of the ecosphere – and nothing more – we should, like every other animal, be occupied exclusively with our own survival, comfort and safety.

(ibid.: 103)

Ecofeminists also argue that the unbalanced exploitation of nature is the result of social domination, but they argue that *all* domination hinges fundamentally around the domination of women by men. Feminists have pointed out that thousands of years of patriarchy have left the self 'gendered'. Ecofeminists argue that the kind of distinction between the human and the non-human world, between humanity and nature, which has placed the non-human beyond the limits of the moral and allowed it to be exploited and used as an instrument for the satisfaction of humans, is parallel to and intimately connected with the moral distinction between men and women which has allowed women to be used as an instrument for the satisfaction of men. Thus, Warren draws together the strands of ecofeminist ethics: 'What all ecofeminists agree about', she writes, 'is the way in which the logic of domination has functioned historically within patriarchy to sustain and justify the twin dominations of women and nature' (1990: 131). The ecofeminist sense of self is constituted by a loving and caring relationship with the non-human world. There is no moral boundary between human and non-human because such boundaries (as between men and women) justify domination.

However, the dissolution of boundaries is itself problematic if it leads to the denial of real difference, as Plumwood (1993) cogently argues. What can easily happen is that the masculine/human moral norm is taken to be the norm for the whole species, or whole ecosystem (Nozick's 'encompassing love', for example). There is something oppressive about an expansion of the self which takes in all non-human species and even all non-human nature. For it is not possible to escape the fact that such a self picture comes from a *human* mind, and as the feminists argue, not coincidentally from the masculine mind. So on this analysis it is not trivial to argue *against* the more extreme versions of ecocentrism on the grounds that human thought is irrevocably bound to the human species.

Moreover, drawing a parallel between 'speciesist' and sexist thinking unwittingly demeans women. Men can learn human moral conduct from women as well as they can from men. But human persons cannot learn human morality from even their closest non-human relatives. This is not to say that animals do not have a morality. They have, but their morality draws very strong species boundaries. Other species may be treated as prey, and within the species, hierarchy is the organising principle, with males usually dominant, and individual competition a genetic norm. Human behaviour which we regard as enlightened is far removed from animal morality.

Following Gilligan's reasoning, Plumwood (1993) argues that the 'rights' base of justice which emphasises separation, individuation, autonomy, reason and abstraction has acquired exaggerated prestige in the public sphere. Whereas:

> a more promising approach for an ethic of nature, and also one much more in line with the current directions in feminism, would be to remove rights from the centre of the moral stage and pay more attention to some other less universalistic moral concepts such as respect, sympathy, care, concern, compassion, gratitude, friendship and responsibility.
>
> (*ibid.*: 173)

But why are these virtues any less universalistic than rights? The point of rights is simply to assert that everyone, regardless of where they are from or the colour of their skin, or how much money they have in their bank account, or whatever their sexuality or gender, is entitled to respect, sympathy, care, concern and so forth. They are *entitled*. Therefore such care, etc. does not depend on whether we happen to feel like giving it.

Important though Plumwood's concepts are in the domain of moral reasoning, their substitution for 'rights' does not resolve the problem of moral difference between humans and the non-human world. The concepts Plumwood mentions are as 'human' as the concept of rights. They apply in nature only to the extent that natural entities approach the human, as do some mammals. Pursuing Plumwood's argument to its logical outcome suggests a different conclusion, namely that different norms relating to the treatment of 'the other' apply in different parts of nature. Killing and eating other species is consistent with the being of predatory mammals. Any other norm would be out of place. Killing and eating *their own* species is not consistent with the being of predatory mammals but is consistent with, say, certain insects.

In an intensely personal essay Florence Krall (1994) explores the natural, philosophical and psychological space of the 'ecotone', the boundary space between homogeneous domains. In ecological terms, the ecotone is a place like the shoreline of a sea or the edge of a forest where it is possible to have something of the best of two worlds. It is possible, she argues, to accept and welcome a space of ambiguity where different values coexist: 'In the natural world, edges where differences come together are the richest of habitats. Animals often choose these ecotones, where contrasting plant communities meet, to raise their young where the greatest variety of cover and food can be found' (*ibid.*: 4). So the boundary between universality and individuality is a space for human growth.

We have been reminded repeatedly that global problems must be resolved locally, yet localities are inhabited by uninfluential individuals whose inequalities cannot be explained away by planetary discourses or international rhetoric. On the other hand, emphasis on differences and distinctions erases the commonality that we seek. However trite it may ring, we are all children of the Earth. Our continuation, no matter where our particular home, what our ideology, or how we make a life, relies fundamentally and inextricably on the health of this planet.

(Krall, 1994: 5)

Krall deploys the metaphor of the ecotone against the demand that we be one thing or the other, that we must always and immediately choose between the poles of some dualism: male or female, them or us, freedom or welfare, justice or care. Living at the creative ecotone means living with contradiction. As to justice to nature, with Stanley Diamond (1981) she commends the: 'primitive' life, the life of the person in the primary group whose environment is a constant part of the group's experience :

The self is seen as an extension of Nature and of the universe. One feels responsible for and related to all things, although valued as a separate and distinct entity. Although a person lives in freedom, a sense of limits is always present.

(Krall, 1994: 225)

How, then, can the global conditions be created which would allow such local communities to flourish? Merchant (1996) begins to approach institutional solutions. She deploys her experience of the Rio Earth Summit to identify three forms of environmental ethics in conflict. The 'egocentric ethic' is inscribed in the GATT, now the World Trade Organization. This ethic embodies one extreme of the bounded individual in which non-human nature is still seen as an infinite resource for exploitation. This is the ethic of entitlement. A 'homocentric' ethic is inscribed in the United Nations Commission on Environment and Development. This ethic also embodies the bounded self but under the political rubric of utilitarianism to which 'sustainability' is added: the greatest good for the greatest number (of humans) for as long as possible. The third ethic is an 'ecocentric' ethic embodying the enlarged self, the ecological self: 'Ecocentrism expands the good of the human community to embrace the good of the biotic community. From an ecocentric point of view, accountability must include the rights of all other organisms, such as those in a rainforest, to continue to exist' (Merchant, 1996: 215). This ethic is inscribed in the institutional network of non-government organisations representing many 'environmentalists'.

Merchant criticises the ecocentric perspective for failing to distinguish between human domination of nature and *capitalist* domination, for failing to recognise the requirements of social justice (what we have termed above 'environmental justice'), and for privileging the whole at the expense of the individual. Against these three competing ethical positions, she counter-poses a fourth which she calls a 'partnership ethic'. Such an ethic incorporates the best principles of both homocentric and ecocentric ethics while rejecting the egocentric ethics of property and entitlement. While not denying the reality of patriarchy, Merchant seeks to sort ethical elements of enlightenment thinking about justice from the Enlightenment's promethe-nian and patriarchal tendency: 'A partnership ethic of earthcare means that both women and men can enter into mutual relationships with each other and the planet independently of gender and does not hold women alone responsible for "cleaning up the mess" made by male-dominated science, technology and capitalism'. A partnership ethic has four precepts:

1 Equity between the human and nonhuman communities,
2 Moral consideration for humans and nonhuman communities,
3 Respect for cultural diversity and biodiversity,
4 Inclusion of women, minorities, and nonhuman nature in the code of ethical accountability (*ibid.*: 217).

A partnership ethic takes humans to be an equal partner with nature, and both as agents. The implications which follow from such partnerships are that practical judgements respectful of nature must be made in specific cases in which the needs of nature and humans conflict. Hence:

> We leave some rivers wild and free and leave some flood plains as wetlands, while using others to fulfill human needs. If we know that forest fires are likely in the Rockies, we do not build cities along forest edges. We limit the extent of development, leave open spaces, plant fire-resistant vegetation, and use tile rather than shake roofs.
>
> (*ibid.*: 221)

Obvious common-sense regulatory judgements perhaps, but judgements opposed by the narrow utilitarian/entitlement basis of justice which domi-nates governance today. The demolition, advocated by feminist critique, of invidious discriminations based around gendered perceptions does not mean the demolition of the ability to discriminate, to make judgement. Ecofeminists have demanded new discriminations and judgements, and a new awareness of contradictory desiderata.

EXPANDING THE SELF

The theorists of 'deep ecology' insist that the self is not a bounded entity. Fox (1984: 194) points out that deep ecology rejects 'discrete entity meta-physics', that is the underlying assumption that what we have to deal with in the world is a collection of items, whether those items be persons, or atoms or cells. In this view the perception of boundaries is illusory. It is not just that 'everything is connected to everything else' but that there *exists* only a single matrix of being, parts of which we *perceive* as entities. These theorists invoke the insights of quantum mechanics and other recent developments in science, as well as those of Eastern mysticism, in support of this perspective (Capra, 1982; Spretnak and Capra, 1986).

Such a conception is problematical for justice and indeed for any morality. Fox (1990) has conducted an examination of the self-pictures of deep ecology and cites a number of writers who regard the development of ecological consciousness as superseding the need for morality (especially Drengson, 1988; Livingston, 1984; Rodman, 1977; and Macy, 1987; cited by Fox, 1990: 227–9). Macy puts the matter in the simplest terms:

> Sermons seldom hinder us from pursuing our self-interest, so we need to be a little more enlightened about what our self-interest is. It would not occur to me, for example, to exhort you to refrain from cutting off your leg. That wouldn't occur to me or to you, because your leg is part of you. Well, so are the trees in the Amazon Basin; they are our external lungs. We are just beginning to wake up to that. We are gradually discovering that we *are* our world.
>
> (Macy, 1987: 20)

A naive understanding of the self as coterminous with the world might conclude that there is no need for moral discriminations and therefore no use for the idea of justice in making such discriminations. There can only be a 'biospherical egalitarianism in which all manifestations of nature have equal value. But of course, as we have discussed above, the inability to discriminate is politically dangerous. Both Fox (1984) and Naess (1984), in his response to Fox, recognise this. Fox says that 'the degree of sentience (of creatures) becomes extremely relevant in terms of how humans relate to the rest of nature if they are to resolve genuine conflicts of value in anything other than a capricious or expedient manner' (1984). Naess (1984) accepts 'certain established ways of justifying different norms dealing with different kinds of living beings'.

In deep ecology we find both a 'realist' rejection of the subjectivist phenomenalism of the post-modernists and post-structuralists, and an absorbtion of eastern mysticism (for example, Zimmerman, 1993a). This aspect of ecocentric thinking has prompted Eckersley (1992) to find something like an unbridgeable gulf between Enlightenment thinking, in which ideas of justice are embedded (and also domination of nature), and the ecocentrism of deep ecology. Zimmerman (1993b; 1994) and Hayward (1994) disagree. Domination of nature, they think, is *contingently* and not *centrally* related to Enlightenment thought. Zimmerman (1993b) points out that deep ecologists, especially Naess, retain a strong commitment to a 'progressive' idea of human evolution. Both Zimmerman and Hayward look forward to a further expansion of consciousness (the self picture) in a dialectical process of critique and debate which carries forward the Enlightenment project.

Hayward shows that Kant represents a kind of Enlightenment thought which is not fundamentally opposed to ecological thinking. We may not be able to characterise Kant's thinking as 'ecocentric', he argues, but the nature of critical reason means that it is wrong to counterpose 'Enlightenment' and 'Ecology' as simple, undifferentiated categories. Thus, 'the Enlightenment is not adequately understood as a dogmatic assertion of the claims of reason over nature, nor should ecology be taken to mean the reverse; rather, they converge on the project of critique' (Hayward, 1994: 53). Hayward sees Kant as representing the 'highest', most developed form of Enlightenment thought and finds much potential within Kant for an ecologically sensitive reason: an 'ecological humanism'. Indeed, he concludes, Enlightenment thought is diverse in important ways. The Enlightenment's 'emancipatory and critical force might actually be required by ecology when applied to the sphere of values' (*ibid.*: 51). Arne Naess, arguably the founder of deep ecology, was strongly influenced by Spinoza, a major Enlightenment thinker, and in order not to misunderstand Naess it is necessary to understand a little more about Spinoza, especially the connection between Spinoza's ontology and his ethics.

In the *Ethics* Spinoza ([1677] 1992) conceived of the universe as a unique all inclusive totality which he calls God or Nature, *Deus sive Natura*. God, (mind) and Nature (substance) are an indivisible whole (cf. Bateson, 1979). Moreover, God or mind is an aspect of this whole, just as Nature or substance extended in space is also an aspect. Hampshire writes of Spinoza's ontology:

> There can be only one substance so defined, and nothing can exist independently of, or distinct from, this substance. Everything which exists must be conceived as an attribute or modification of, or as in some way inherent in, this single substance. This substance is therefore to be identified with Nature conceived as an intelligible whole.
>
> (Hampshire, 1951: 31)

Both the mind–body dualism and the human–nature dualism are therefore absolutely rejected.

Spinoza then proceeded to use this ontology to identify the nature of moral or ethical conduct. Spinoza says:

> Man knows himself only through the affections of his body and their ideas (remembering that mind and body are aspects of a whole). When therefore it happens that the mind can regard its own self, by that very fact it is assumed to pass to a state of greater perfection, that is, to be affected with pleasure, and the more so the more distinctly it is able to imagine itself and its power of activity.
>
> (Spinoza, [1667] 1992, *Proof of Proposition* 53)

This pleasure stems from the innate tendency in the human being (as in all beings) to sustain itself (the 'conatus'). It is natural to seek to perfect ourselves and thus to come closer to realising the fullness and richness of which we are capable. Pain, on the other hand is the feeling of impotence, the feeling that we are at the mercy of forces both external and internal, and that we are not the author of our actions. We are passive; we suffer. A passive emotion (one from which we suffer) 'ceases to be a passive emotion as soon as we form a clear and distinct idea of it' (*ibid*. Prop. 3, Part V). Pain, 'is man's transition from a state of greater perfection to a state of less perfection'.

Self-knowledge, and through it self-realisation, cannot be accomplished without an understanding which views the self in its *complete environment*, which is God or Nature. Feldman observes: 'Thus on Spinoza's view, what makes a person an agent is self-knowledge; lacking such knowledge, an individual is merely a passive (suffering) recipient of external and internal stimuli to which he responds either blindly or inadequately' (1992: 15). Our freedom consists ultimately in behaviour which is in accordance with the self-realisation of God or Nature. Self-realisation through understanding our own place in Nature necessarily entails abandoning an anthropocentric viewpoint and rejecting the idea that everything in Nature is a means to the advantage of humans, as Spinoza's *Appendix* to Part I makes clear:

> Looking on things as means, they (men) could not believe them to be self-created, but on the analogy of the means which they were accustomed to produce for themselves, they were bound to conclude that there was some governor or governors of Nature, endowed with human freedom, who have attended to their needs and made everything for their use. Thus they worship god in such a way that 'he' (the god) will direct the whole of Nature so as to serve his (man's) blind cupidity and insatiable greed. . . . But in seeking to show that Nature does nothing in vain – that is, nothing that is not to man's advantage – they seem to have shown this, that Nature and the gods are as crazy as mankind.
>
> (Spinoza, [1677] 1992)

Naess, like Spinoza, uses rationalist tools in his philosophy. But, as Kant pointed out, in the human condition rationalist logic meets its limits. More than other transpersonal ecologists, Naess confronts the contradictions in the human condition. When we understand that when we harm others we harm ourselves, he argues, non-instrumental acts become instrumental. Yet he explicitly extends Kant's categorical imperative: 'You shall never use any living being only as a means' (Naess, 1989: 174). Every life form, including humans, has the right to self-realisation – the full flowering of its potential as a species. But the self-realisation of one life form can be the destruction of another. 'Equal right to unfold potentials as a principle is not a practical norm about equal conduct towards all life forms. It suggests a guideline limiting killing, and more generally limiting obstruction of the unfolding of potentialities in others' (*ibid.*: 167).

In Naess's view, there can be no absolute rule for *practical* use. The creation of practical norms requires a discursive process by means of which we humans (in accordance with our humanness) work out practical rules (make practical judgements) for different life situations (*ibid.*: 168). Applying this thinking, we might find different answers to the question of whether it is right to kill an animal in different situations, for example: when the animal is threatening to kill a human, when the animal is damaging a human, when a human would otherwise starve, when a human would otherwise be undernourished, when killing an animal (or conducting laboratory experiments) would save or prolong human lives, when killing an animal would meet a preference of another human (hunting for sport, or making cosmetic products from animals). Even if, in some circumstances, we may judge it right to kill an animal, a different judgement might well apply to the mass cultivation and commodification of animals to satisfy human tastes for food.

Spinoza, at the beginning of the age of science and rationalism, like Descartes, believed that all apparent contradictions could be resolved by logic. The mind which produced logic was an aspect of Nature. Therefore in so far as nature resolved its contradictions, so too must logic. Spinoza's mistake was to coflate mind with logic (see Bateson, 1979: 58). Naess relies upon and trusts his intuiton. With his increased awareness of the limitations of logic, he knows that he will not succeed in resolving human contradictions by reason alone. 'The theoretical starting points of the philosophy of the one and the many cannot replace the concrete time and situation determined deliberations which must be made in a choice of appropriate political action' (Naess, 1989: 195). 'The complete formulation of an ecosophy is out of the question' (*ibid.*: 196). But the struggle to resolve contradictions must continue both within the person and within society by means of such tools as are available. Principles and logical arguments are necessary if thoughtful judgements are to be made. But the practice of judgement cannot be reduced to a single logically consistent set of principles.

Naess, nevertheless, offers the following formulation of (human) environmental justice (1989: 207). First, mere wishes of humans have to be distinguished from needs. Basic biological needs, of course, have to be satisfied for an individual or a species to survive. But, second, the base-line is not survival but 'minimum conditions for Self-realization', that is: minimum satisfaction of biological, environmental and social needs. These conditions should have priority before others. Third, 'under present conditions many individuals and collectivities have unsatisfied biological, environmental and social needs, whereas others live in abundance'. Therefore, fourth, 'To the extent that it is objectively possible, resources now used for keeping some at a considerably higher level than the minimum should be relocated so as to maximally and permanently reduce the number of those living at or below the minimum level'.

Mathews's theory of value, stemming from the Spinozan idea of 'conatus', resonates with Benton's ecological continuum but enables some important further discriminations to be made (Mathews, 1991). Conatus is the impulse to self-maintenance and self-realisation (identical in Mathews's view). Conatus is identical with the essence of the being of a thing. However, where Spinoza allowed everything that exists its conatus, Mathews, following a systems-theoretic line, includes only organic life forms. An organism 'is first and foremost, a system for effecting its own maintenance, repair and survival' (*ibid.*: 98). Organisms are intrinsically self-realising. 'The categorical difference between organisms and other individuals – systems or substances – is that organisms, by their activity, define an interest, or a value, namely their own' (*ibid.*: 100). Mathews solves in a pragmatically satisfying way the problem of where 'values' come from. Organic life forms value themselves.

Intrinsic value, thus, lies in an organism's capacity to define itself from within and not have it imposed from an external source. Mathews avoids the term 'life', but that does seem to be what she is talking about.

> What the thesis principally points to is that with the appearance of self-realizing systems, something new, something qualitatively, indeed categorically, different from the rocks and clods of clay and grains of sand that populate the class of substances, enters the world.
>
> (*ibid.*: 104)

The rocks and clay possess only a contingent individuality shaped by external forces: 'a rock is only an individual by chance, and its individuality does not matter to itself' (*ibid.*).

It is important here to distinguish between human consciousness and the wider concept of 'mind'. On the one hand, the concept of 'mind' concerns the selective process which is to be found throughout nature: 'Primarily living forms react to external stimulation in such fashion as to preserve the living process. The peculiar method that distinguishes their reactions from the motions of inanimate objects is that of selection' (Mead, [1932] 1980: 71). Bateson (1979: 126–7) concludes that both *autonomy* and *death* are characteristic of mind, a capacity for purpose and choice through self-corrective processes, the capacity to learn and remember, and the capacity to unite with larger systems to make larger wholes. None of these characteristics are limited to the human being. Consciousness, on the other hand, arises when the living creature is able to respond, as Mead ([1932] 1980) puts it, to its own responses. This means that the creature becomes aware of its responses to its environment, and therefore, in a sense, the boundary between the subject and its environment becomes unclear. As E. Murphy says in his introduction to Mead's book:

The relations in which the environment stands to our reactions are its meanings. To respond to such meanings, to treat them, rather than mere immediate data, as the stimuli for behavior, is to have imported into the world as experienced the promise of the future and the lesson of the past.

(Murphy in Mead, *ibid.*: xxxiii).

Mathews (1991) posits three levels of 'intrinsic value'. Each more basic level subsumes the less basic. The first is inherent in all things because all things are part of a universe which is arguably a 'self' (or 'mind') in the sense of being a self-realising whole. This does not take us far. It is a background 'value'. The second level is the value each organic life form has on account of its capacity for self-realisation. The third level is the value every object (including life forms) has for every other life form. The third level is therefore entirely relative to the interests of each life form, including that of humans.

The second level of value is the most significant for, 'it is only the second level of value that furnishes normative indicators. The second level of value is identifiable as the intrinsic value embodied in individual selves or self-maintaining systems' (*ibid.*: 118). Intrinsic value entails 'respect'. Mathews therefore explicitly extends the Kantian 'kingdom of ends' to include the non-human world (*ibid.*: 119). The field of second-level values is further differentiated according to the degree of self-realisation of which an organism is capable and this is in turn dependent on the organism's level of complexity (*ibid.*: 123). This postulate leads to the position that beings that can maintain themselves as individuals (such as blue whales) have greater intrinsic value than beings which can only maintain themselves either in close interaction with an environment (amoebae) or in close interaction with one another in a community (ants). However, this position is complicated by ecological interdependencies. A simple continuum based on individuality is out of the question. The intrinsic value of a given self is 'a dual function of its relative autonomy, or power of self-maintenance, on the one hand, and of its interconnectedness, or dependence on other selves, on the other hand' (*ibid.* 126). Neither aspect of value can be wholly reduced to the other. Nevertheless, within this logic an individual member of a species of complex entities such as blue whales has a higher value than an individual member of a species of much less complex entities such as krill which blue whales eat. But it makes no sense to assign a higher value to blue whales (as a species) than to the species of krill on which they depend.

The human person is different again from the species of animals. The key difference for Mathews is our ability to 'grasp our unity with the wider whole'. Specifically *human consciousness* is a defining feature of the human being. Self-realisation takes on an additional meaning, namely the capacity to seek identification with wider wholes – the 'ecological self'. This is not to be understood as a contingent feature. Value does not derive from *how* conscious any of us actually are or *how* well we may have achieved this kind of self-realisation. Intrinsic value derives from the intrinsic capacity, the conatus, of the human being to achieve that particular kind of consciousness: 'When we recognise the involvement of wider wholes in our identity, an expansion in the scope of our identity and hence in the scope of our self-love occurs' which leads to 'a loving and protective attitude to the world' (*ibid.*: 149). Mathews does not draw the conclusion from this difference that humans have a higher intrinsic value than other animals. That is established, if at all, by the degree of autonomy and complexity of the human species.

RETHINKING THE BASES OF JUSTICE

'Ecological consciousness' is an expression of the human desire for an expanded conception of the self. Conceptualising the transition from being to action should be the primary philosophical task of the movement today. The potential of an expanded conception of the self to lead to an enlarged conception of justice must not be subverted by extremist distortions of the relationship between ontology and action. An expanded conception of the self does not dispose of the need to make practical discriminations. Ecological consciousness requires an expanded and not a diminished sense of justice.

All species, not only humans, have the potential to destroy the richness and diversity of their local environments. But, because of their dependence on that environment, its destruction eventually leads to destruction of the species in sufficient quantity to restore the balance in the long run. But humans have developed in such a way that two imponderable new aspects have begun to show up in their relationship to the ecosystem they inhabit. First, the human species is a truly global species and its activities have begun to affect not just a part but the whole of the ecosystem (albeit bit by bit). Second, humans have been able to outwit, to some extent, the retroactive impact of local environments on the species to enable it to survive massive environmental changes and avoid species destruction. These tendencies raise the possibility that, in the long run, the ecosystem as a whole will exert its retroactive impact on the human species *as a whole*, bringing about massive human destruction. A concomitant of such an event might also be reduction in the richness and variety of the ecosystem as a whole. An alternative is for the human species further to develop its normative systems to avoid that outcome.

Such further moral development requires moral discriminations similar to those of Benton, Merchant and Mathews. In our view, the relationship between humanity and nature is best described as asymmetrically co-dependent. We can appreciate today that the survival of the natural world is dependent upon what humanity does. At the same time humanity remains completely dependent for survival upon non-human nature, that is to say

155

upon our planetary biosphere and all its inhabitants. Of course, there are individuals and societies in both nature and humankind; but there is still an important moral distinction between human and non-human 'nature'. Whereas much of non-human nature may be thought of morally in terms of species and ecological systems, humanity *must* be considered morally in terms of individuals, even if we also recognise the dependence of individuals on human communities. We have seen no argument to refute that Kantian principle. A lesser principle appears to us to apply to animals. If we want to be just towards other humans, we have to accept that sometimes animals may be sacrificed for the good of humans.

We should try to go further than Merchant in specifying the practical implications of unequal asymmetrical co-dependence. We agree with Merchant that many issues must be subject to judgement taking into account the specific situation. However, not to posit some general principles is to avoid a large part of the issue. A judge must be able to draw on some principles in deciding a case. If there are to be adequate institutional means for the definition and delivery of justice in and to the environment, as we argue in Chapter 7, we would expect these principles to be much elaborated and modified in practice.

Based on the proposition that the human self reaches its highest form of expression in connection with the natural world, we therefore propose the following:

The first principle of ecological justice is that every natural entity is entitled to enjoy the fullness of its own form of life. Non-human nature is entitled to moral consideration. With an extended conception of the self, an absolute barrier between human and non-human nature is untenable.

The second principle is that all life forms are mutually dependent and dependent on non-life forms. This principle must be considered when any conflict among species occurs. Exactly what implications this principle has for judgement in specific instances of conflict between the rights and needs of different life forms is as yet unclear.

Subject to the above principles, we make three distinctions. These are moral 'rules of thumb' which we think are broadly consistent with Mathews's argument.

1 Life has moral precedence over non-life.

> Non-living nature takes value from its function as part of the habitat of living nature. We include bacterial and plant life, though viral life and machine life (sophisticated 'thinking' computers) are borderline. Mountains and streams, for example, take their value from the life, including ecological systems, they support. Were the Earth threatened with the collision of an

asteroid, humans would be entitled to deflect it from the Earth by any means whatsoever.

2 Individualised life forms have moral precedence over life forms which *only* exist as communities.

> There is undoubtedly a continuum here between a community life-form, such as an ant colony or bee swarm, and a form in which individuals can become highly differentiated as in a pod of dolphins, a herd of elephants, or a community of apes. This distinction would warrant human intervention to protect a colony of individualised animals from attack by, say, bacteria or insects.

3 Individualised life forms with human consciouness have moral precedence over other life forms.

> An animal may be killed if it is attacking a human. Higher animals may be sacrificed for food under well-regulated conditions which respect their rights to a full existence. An elephant which is destroying the vegetable plot on which a human family depends (as happens in India from time to time) may be killed if it cannot be prevented from intruding by other means.

The categories are fuzzy and the boundaries between them indistinct. They are not meant to become unchallenged rules. Such distinctions are always debatable. All life forms deserve certain rights to the fullness of their natural existence but a biospherical egalitarianism cannot be sustained logically or practically. And if we are going to make distinctions in practice, we had better make them explicit so that that they can be debated. The most contentious distinction is that between human and non-human life. The argument in favour of this distinction (which has always been defended only as an axiom) is that it should stand until it is adequately refuted. This is an expression of the precautionary principle applied to politics. It must be refuted not just in terms of deontology, but consequentially. We must be satisfied that the political and social consequences of removing it are acceptable to humans. If that criterion cannot be met, then if it is applied, the ethics of discourse will be breached and, in any case, the norm is unlikely to be sustainable over a long period in any human society.

CONCLUSION

The ideas discussed above seem to us to indicate valid paths for opening up 'justice' to non-human nature. We did not expect, and nor did we find, consensus. The debates about ecological justice are consistent with our view that finding justice is a dialectical process. The question which we must now confront is whether the dialectic can be enlarged in practice to include this field, whether institutions can be created which will enable the debates to have some practical impact in the world. We have to consider whether, and if so how, ecological justice can become a moral goal before the material fact of our mutual dependency on each other and on nature forces itself upon us in a series of catastrophic human and ecological crises.

Ecosocialists seem disposed to be optimistic about the possibility of institutional change. But it is possible that humanity will *not* be able to extend its morality in time to prevent catastrophe. Class antagonisms may obscure the fact of mutual dependency. The extinction of species started to occur long before humanity began to influence the rate of extinction. Humanity's own destination may well be self-extinction. Whether that will be the case probably depends on *when* rather than *if* the mutual dependency of society and nature is fully understood by enough people with the power to change human society and human understanding.

Human society has certainly been transformed over the last two hundred years but hardly as a result of human intention. If we look fearlessly at our human situation today, a pessimistic outlook seems more honest (see Bailey, 1988). However, if the above principles of ecological justice are to be taken forward and tested in practice, if they are to be given some real chance of having an impact on our relationship with nature, then we must consider the justice of the institutional conditions of our world today. To this we turn in the next chapter.

7

JUSTICE AND NATURE

New constitutions?

INTRODUCTION

An international political system, a global network of institutions, has been taking shape for the last three hundred years or so. The institutional system mediates the delivery of justice. The question we want to address in this chapter is simple: is this system itself environmentally and ecologically just?

We view this network of institutions fundamentally as a political system. Such a view follows from a primary concern with justice. Nevertheless, the system is both a system of production and a system of governance. The system of production works only partly, as Hayek puts it, as a *catallactic* order: a self-organising, self-governing system. It works also because there is a nexus of competing nation states loosely submitting to global institutions of governance: for example, the World Trade Organization, the United Nations.

The international system has expanded spatially from a European base over the last three hundred years. In the course of its expansion it has encountered resistance from other systems – of production and governance, materiality and ideas. But it does not yet extend over the whole globe. Indeed a crucial question today concerns who is and is not going to be part of it; who is included and who excluded from its benefits and costs. Nor does the international system yet regulate all aspects of life, even in those territories in which it holds maximum sway. One part of life which the system regulates only partially is its own natural environment. The natural environment offers a new and resilient resistance to the further expansion of the system of production.

What we want to bring to light is what this international system is like both materially and ideally, as a system of production and a system of governance. In the first section we consider ideas about the future of the international system as a system of production, in the second its future as a system of governance. We draw on the recent work of scholars who have themselves investigated the international system. We consider the justice of these two aspects of the international system in the light of the expanded perceptions of morality discussed in the preceding chapter, and our under-standing of justice as a dialectical process. In the final section we sum up our observations of the environmental and ecological justice of the political system and consider what are the 'next steps' to be taken to move this system in a progressive direction.

THE INTERNATIONAL POLITICAL ECONOMY AS A SYSTEM OF PRODUCTION

There are many political–economic perspectives on ecology, several of which overlap on key aspects. For example, many writers contributing separately to sustainable development and ecological modernisation discourses share similar views on the questions of market regulation and economic growth. Our purpose here is not to discuss an exhaustive typology of perspectives, but rather to illustrate important divergences of opinion on the question of capitalism's future. To this end we identify and briefly discuss three broad groupings of contrasting opinion: market environmentalism, ecological modernisation and ecosocialism. (For a more comprehensive review of envi-ronmental economics perspectives, see Eckersley, 1995; Jacobs, 1995; and Turner, 1994.)

Market Environmentalism

Since the 1960s, an increasing number of conventional economists have attempted to reconcile market theory with the evidently worsening problems of ecological pollution and resource depletion. These analyses have converged around certain key assumptions which define the discourse of 'market envi-ronmentalism' (ME). We include under this common heading the distinct but closely related environmental analyses of welfare–utilitarian economists and public choice theorists. Recalling our analysis in Chapter 4, it is important to draw attention to the fact that these two perspectives provide different justifi-cations for the existence of key institutions, such as market exchange, the state and private property. For example, public choice analysts share many of the assumptions of entitlement theorists, including a hostility to taxation, or any other redistributive mechanism which infringes property rights. By contrast, welfare–utilitarian economists recognise the inevitability of certain market failures – notably externalities – that necessitate corrective state action, such as polluter taxes which enforce the internalisation of environmental costs by economic units (see Eckersley, 1995; and Jacobs, 1995, for finer-grained anal-yses of the market perspective).

In spite of these important theoretical divergences, both public choice and welfarist environmental economics share key theoretical and political assumptions which suggest the common grouping we have assigned here, under the 'ME' rubric. First, as discussed in Chapter 4, both theories share certain (in our view, problematic) epistemological features, including individualism, monism and anthropocentrism. Second, both theories view the market as the key mechanism of social and ecological integration. Third, the perspectives tend to support a similar, if not uniform, political programme involving both deregulation of all direct state intevention mechanisms and the extension of market relations to all aspects of environment and society.

Some of these broad assumptions of ME have been usefully summarised by Seldon:

> The consumption of the environment can be analysed by economists in the same way as commodities and services in general. The environment – pure air, clean water and so on – is a scarce resource that is used in the production of goods and services by industry, public utilities, nationalised industries, local and central government. It must therefore be 'economised', so that it is only used to the point at which its social costs are covered by the social benefits. And this of course is equally true of scarce labour, equipment and capital used in production. The question is whether industry or government can be induced to economise its use more effectively by charges than by direct regulation.
>
> (Seldon, 1990: 6)

In recent years ME has been codified in a series of authoritative and influential collections (e.g. Bromley, 1995) and reviews (e.g. Cropper and Oates, 1992).

An influential early advocate of ME, Beckerman (1974; [1975] 1990; 1994), argued that humanity's use of the environment was better disciplined by market prices in the form of charges than by direct government control. The market pricing approach of Beckerman and other welfare–utilitarian economists has been followed in recent years by an important strand of ME that has undertaken a thoroughgoing critique of 'command and control' (i.e. state interventionist) environmental policies. Reflecting many of the assumptions of entitlement theory, these public choice analyses have sought to emphasise the ecological superiority of completely unfettered markets over any form of regulation, including the resource and pollution taxes advocated by Beckerman and other welfare economists (see Eckersley, 1995).

A central assumption of ME is that environmental problems are really only resource misallocations, and arise mainly because of the absence of clear property rights, thus denying nature the efficient allocative logic of the market. Indeed, the arguments for assigning ownership rights (emphasised by public choice theorists) and user pricing (emphasised by welfarists) to all aspects of the environment have become the central nostrum of ME (e.g. see Brubaker, 1995; Simon and Kahn, 1984). It is assumed that conferring property rights on common or socially owned resources will draw nature itself into the spheres of commodity production and circulation. Given that nature is regarded as simply another factor of production, ME argues that its subsumption into market realms will ensure the efficient allocation of environmental values and thus prevent both the exhaustion of resources and the ruin of ecological systems.

The commodification of nature requires its monetisation, and every conceivable value must have a price if exchange is to take place resulting in efficient distribution of values. Thus, a vast literature has emerged on the various methods needed to place money values on nature in order to permit the pricing of environmental 'goods and services' (see review by Bateman and Turner, 1994). These methods have been vigorously contested. An equally extensive literature has also arisen to refute the attempts of ME theorists to price ecological values – criticisms have centred variously on the perceived technical, moral and political failings of valuation methods (Turner, 1994).

As the main ecological face of the powerful discipline of economics, ME has conditioned policy development both by nations and by the guardians of the new globalising market economy, such as the World Bank, the OECD and the European Union, and the World Trade Organization (see Eckersley, 1995). The ME doctrine has been promoted by academic economists and, particularly in western countries, by a range of neo-liberal lobby fora, such as self-described 'think tanks' and private research institutes (e.g. Anderson and Leal, 1991; Bennett and Block, 1991). Thus, the application of property rights to all parts of nature, a policy ceaselessly advocated by ME theorists, can be regarded as a key aspect of globalisation, the contemporary process whereby market relations are both expanding within capitalist nations and extending to non-capitalist countries and communities.

As Seldon observes, a fundamental assumption of Beckerman's welfarist analysis is that 'it is wrong to regard the environment as an absolute that must be preserved at all costs' (1990: 8). In Seldon's view, 'Beckerman cogently demonstrates that it is appropriate to use the environment in the course of production if the loss of environment is exceeded by the gain in production of goods and services' (*ibid.*).

Importantly, ME is both resolutely anthropocentric, taking no regard for the argument that nature has intrinsic worth, and firmly instrumental, viewing nature as a source of human gratification (utility). Beckerman's (1994) attack on the ideal of sustainable development is a definitive example of anthropocentrism. Clearly, if the consumption of nature is to be regulated in accordance with social costs and benefits, the question of non-human values cannot enter the spheres of political governance. Moreover, this instrumental calculus permits almost any use (and abuse) of nature by particular human communities, providing they can tolerate the real and perceived outcomes of such practices. ME hardly problematises the ideal of economic growth in any profound sense; indeed, Beckerman's (1974) 'polemic' stoutly defends it. The perspective conceivably permits any rate of resource consumption if the sum of individual utilities can outweigh the aggregate costs to humans of growth. The future is discounted, thus jeopardising future generations. As a consequence, higher rates of economic growth become easier to justify in contexts where people and institutions place a low value on environmental quality. Moreover, as was pointed out in Chapter 4, the difficulties of measuring and aggregating individual values renders such calculus both inherently unreliable and open to political manipulation.

Of course, it should not be surprising that ME theorists vigorously defend growth, for to refute this ideal is to throw doubt on the future of capitalism itself. Capitalism's central dynamic, even leitmotiv, is the process of valorisation, the ceaseless expansion and accumulation of value. Without expansion, the process of accumulation must fail. Ecosocialists argue that the self-valorising logic of capitalism goes to the very heart of the ecological crisis.

The fact that cultural and economic values are individually distributed makes the notions of 'social' costs and benefits inherently elusive, if not illusory. ME theorists have considered the distributional consequences for individuals and firms of imposing allocative efficiency on resource use (say, through the introduction of market pricing of resources and pollution taxes). Most remedial strategies involve familiar utilitarian compensation and mitigation principles (e.g. Beckerman, [1975] 1990: 71–7). Ultimately, however, the usefulness of such remedial policies is severely limited both by the conceptual and political priority given to efficiency as the key decision criterion, and by the need for complex administrative and regulatory systems which can apply and monitor such redistributive measures. In other instances, the policy of economic growth is assumed, by proxy, to address in the medium to long run any perverse distributional effects of market relations, including the exchange of ecological values. Beckerman, for instance, observes that: 'it may well be that faster growth is a necessary condition for shifting the income distribution in a more egalitarian direction' (*ibid.*: 32). Advocates of the growth-redistribution model seem cheerfully (or willfully) ignorant of the repeated historical failures of 'trickle down' economics (Pepper, 1993).

Plainly, also, ME depends upon a view of the bounded and atomised self and cannot accommodate a perspective in which individual selves are defined against the wider horizons of the human community and society of which they are part, let alone against the rest of the ecological system. Taken to its logical conclusion, society dissapears; so too does the planetary or any other ecology. As Hayward (1994: 104) properly concludes, 'ecology into economy won't go'. As we have seen in Chapters 4 and 6 this reduction is not a feature of Enlightenment thought so much as its particular interpretation in the limited sphere of utilitarianism and entitlement theory, which form the narrow ethical basis of modern economics.

In one sense, ME defends the status quo, the globalising institution of the market, and resists the notion that any fundamental structural change is needed. But in another sense, the perspective cannot be seen as promoting 'business as usual', given the insistence upon both the extension of property rights to the whole of nature and the intensification of commodity relations in general. Beckerman (1974), for example, notes the critical role that the state will need to play in creating market-pricing systems which will regulate the consumption of nature. Even the extensification of property rights advocated by the public choice strand of ME will require an enormous administrative effort by states in order that resource ownership might be properly codified and upheld in law. Such policies are hardly conservative – the privatisation of commonly pooled and publicly owned natural resources has proved to be politically controversial in many national and regional contexts.

Thus, ME theorists advocate serious political change, namely a further shift to the market as the solution to ecological problems. However, this view of change is primarily political–economic rather than political–institutional – ME theorists would doubtless be wary of any proposal to build new, or strengthen existing, international institutions for the purpose of ecological regulation. This perspective really only supports extending and strengthening the institutions and codes of formal (legal) justice needed to instate and police the contractual basis of globalising capitalism. While the contractual relations of international commerce might certainly be amended to include environmental considerations, there is no justification for increasing the directive powers of global political institutions, such as the United Nations. In short, if the environment is to be saved, then this must be through the mechanism of the market rather than through the political and regulatory activities of a global authority. This position contrasts with the two other positions to be discussed, both of which view the market as the source of environmental problems, though to varying degrees.

Ecological Modernisation

Market environmentalism tends to deny two arguments that have been broadly accepted in other environmental perspectives: first, that global ecological problems have reached a crisis point, to the extent that all human and non-human life is increasingly imperilled; and second, that the market as currently structured is at least a contributing cause of this crisis. One broad approach which accepts these arguments has been termed 'ecological modernisation' (EM). Blowers (1996) traces ecological modernisation theory to the work of the German sociologist, Joseph Huber, in the 1980s. The persective is heavily influenced by European authors, notably Hajer (e.g. 1995), Janicke (e.g. 1990), Mol (1995; 1996) and Weale (e.g. 1992).

According to Blowers (1996: 3), the EM perspective, 'holds that while environmental constraints must be taken fully into account, they can be accommodated by changes in production processes and institutional adaptation'. Blowers argues that EM 'regards the environmental challenge not as a crisis but as an opportunity'. However, this may be merely a question of emphasis. Unlike market environmentalists, many EM theorists, such as Weale (1992), certainly both make explicit reference to the severity of the 'environmental challenge' facing the globe, and recognise that significant institutional and technological changes are required in order to prevent ecological catastrophe. Moreover, some proponents of EM, such as Mol (1996), hinge their analyses on the notion of a global environmental crisis.

EM problematises the present direction of capitalist development, though its exponents argue that capitalism's ecologically destructive course can be corrected through institutional change: 'encouraged by a market economy and facilitated by an enabling state . . . ' (Blowers, 1996: 3). In short, EM is a reformist perspective which, while recognising the ecological dangers posed by unfettered markets, believes in the self-corrective potential of capitalist modernisation. As Blowers puts it, EM sees the present historical juncture – the ecological crisis – as a moment of *transition*, to a more sustainable modernity, while radical perspectives, such as ecosocialism and deep ecology, view the contemporary as a moment of *transformation* from which will emerge a new social formation.

A central assumption of most EM theorists is that economic growth can be reconciled to the realities of ecological sustainability. For most, this reconciliation can be achieved through the adoption of three main strategies: the ecologisation of production (i.e. the reduction of waste and pollution through technological improvements); the refinement of markets and regulatory frameworks to better reflect ecological priorities; and the 'greening' of social and corporate values and practices (Blowers, 1996). EM theorists are divided on the extent of state intervention needed to achieve sustainable modernisation. Weale (1992), for example, envisages the need for a highly interventionist state, while other EM theorists advocate an 'enabling' model of governance that closely reflects the minimalist state form promoted by neo-liberals (Blowers, 1996).

As Dobson (1990) observes, many EM theorists go beyond the argument that economic expansion can be decoupled from environmental degradation to insist that growth is actually necessary to ecological improvement. Reversing the equation, modernisers such as Weale envisage environmental protection as a source of economic growth, although this vision hinges upon the assumption that environmental amenity can be counted as a superior good which expands the demand for pollution control. However, as Dobson (*ibid.*) argues, the superior good assumption can hardly be applied to the many countries where social values place a weak emphasis on environmental amenity (over, say, economic survival) and/or where environmental regulation is feeble or non-existent (e.g. Papua New Guinea and the Ok Tedi dispute – see Chapter 1). Nor can it be assumed that environmental protection measures assist growth, even in western countries. Recent evidence from Holland (which Weale, 1992, acknowledges) suggests that environmental regulation may depress national output (Dobson, 1990).

Indeed, does the EM model confront the complex problems raised by the uneven globalisation of capitalist political–economic forms? It appears not. Blowers, for example, describes it as 'a theory based entirely on Western industrial experience' (1996: 14). Christoff argues that EM theory demonstrates little appreciation of the complexities of recent and contemporary global changes. According to him, EM theorists:

> offer only a diminished recognition of the increasingly internationalised flows of material resources, manufactured components and goods, information and waste; of the influence of multinational corporations on investment, national industrial development and the regulatory capacities of the nation-state; and of international deregulatory developments (such as GATT) and environmental treaties (such as the Montreal Protocol).
>
> (Christoff, 1996: 486)

As Hirst and Thompson (1996) observe, globalisation has occurred irregularly in time and space and has not produced a homogenised form of western industrialism in every country. Given the diversity of socio-cultural and ecological contexts, there is a glaring inadequacy in any environmental theory which seeks to project on to the globe the experience of any one nation or region. Thus, Blowers observes: 'Subsistence economies which are prevalent over much of the Third World may actually be more sustainable than modern agricultural systems based on the intensification of production' (1996: 14).

Just as critically, Dobson (1990) refutes the EM argument that declining energy consumption per unit of GNP in OECD countries signals a decoupling of growth and environmental depletion. As Dobson notes, this consumption decline can be explained as the consequence of three historically – and geographically – specific factors, two of which cannot be reproduced worldwide. The decoupling of growth and energy use 'was encouraged by high energy prices, faster economic growth of the service sector, and the relocation of energy intensive industries to developing countries' (World Resource Institute, cited in Dobson, 1990: 208).

Christoff's assessment is that, 'certain improvements in environmental conditions in the First World have been gained through displacement of high energy consuming and/or polluting industries (for example, metal processing and primary manufacturing) to newly industrialising countries . . . and lesser developed countries' (Christoff, 1996: 479). Thus, EM may only succeed in particular, *developed*, national contexts, and the costs may include further environmental degradation and resource depletion in other countries and regions.

Christoff's (1996) thoroughgoing review draws attention to the analytical and normative diversity of EM perspectives. As one possible conceptual frame, Christoff envisages a normative continuum of positions ranging from 'strong' EM, emphasising the need for broad social and institutional shifts to prevent and correct environmental degradation, to 'weak' EM, reflecting a narrower focus on technological and market-based solutions to ecological problems. Christoff's continuum echoes the distinction that has been made between 'weak' and 'strong' versions of sustainability (cf. Blowers, 1996). Christoff's definition of EM is more expansive than that proposed by Blowers. For the latter, EM is: 'essentially a conservative theory espousing a weak version of sustainability achievable through a greater emphasis on environmental conservation. It breaks with the idea that environmental needs are in conflict with economic demands' (Blowers, 1996: 12).

Later in the same essay, Blowers describes EM as, 'a celebration of capitalism with a greener face' (*ibid.*: 14). By contrast, Christoff's version of EM accommodates 'strong' perspectives that emphasise ecological over economic priorities and stress communicative, democratic strategies over technocratic, instrumental policies. Thus, Christoff seems prepared to place Beck's work at the strong end of his normative continuum, while Blowers explicitly positions the 'risk society' thesis outside the EM perspective, as a 'radical, transformative' strategy for change.

Christoff's is arguably the more accurate categorisation of Beck's work, given the latter's neo-Kantian faith in the immanent potential of 'reflexive modernisation' to correct the destructive tendencies of modern industrialism. Here Beck echoes the emphases given by certain other EM theorists to the self-corrective rationality of modern capitalism – Mol, for example, speaks of 'reflexive modernity' and believes that 'ecological modernisation can be aligned with systemic or institutional reflexivity' (Mol, 1996: 318). Lash (1993) also places Beck within the modernist theoretical framework. While Beck may be deeply critical of contemporary industrialism, his limited normative sketches seem both to predict and advocate the emergence of a new ecologically sustainable capitalist modernity.

Is Christoff right in his inclusive depiction of EM? Is EM anything more than a 'moderate and conservative theory confirming business as usual'? (Blowers, 1996: 14). If 'strong' versions of EM do in fact present profound criticisms of industrial capitalism, can they still be placed within the 'modernisation' frame? Many strong versions of EM seem equivocal in political–economic terms, given that the premises of such analyses frequently contradict the idea that capitalist modernity can be reformed on ecological principles.

For instance, the systemic adoption of the precautionary principle would heighten immeasurably the 'risk' inherent in most entrepreneurial investments, thus surely undermining the process of valorisation and thereby capitalism. Christoff argues that 'the most radical use of ecological modernisation would involve its deployment against industrial modernisation itself'

(1996: 491). However, as Christoff notes, this is to attack one of the very defining features of modernity (cf. Giddens, 1990). If industrialism is swept away, something different from 'modernity' will surely remain. Moreover, as the 'socialist' experiments of the twentieth century have shown so dramatically, markets and accumulation are indispensable aspects of modernisation. To abandon or even problematise any of the key institutional pillars of modernity may be to advocate a new social form altogether.

In summary then, EM is clearly an inadequate political ecology in the era of global environmental crisis. As Blowers succinctly puts it, EM, 'abstains from a broader diagnosis of the conflicts within capitalist societies, the problem of inequality (especially between North and South) and the trends associated with those industrial processes which, if not arrested, may eventually threaten survival' (1996: 21–2). Moreover, the 'strong' versions of EM that Christoff (1996) identifies, – i.e. those which problematise capitalist modernity on structural grounds and advocate transformative change – may well be better categorised outside the modernisation frame.

A more expansive 'self' picture is accommodated by EM than by ME, allowing the entry, for example, of justice as meeting need. Market 'failure' has been generally regarded as the failure to meet the needs of humans – though it surely does not have to be so restricted. If a wider self-picture can be accommodated within EM, none of its exponents seem explicitly to have addressed the possibility. The definition of 'deep' ecology was supposed to distinguish the reformers who fall into the EM camp ('shallow') from those whose self-picture was more inclusive of the non-human world. But this distinction may itself be specious (see, for example, Bookchin, 1995b: 90–3).

None the less, even the weaker versions of EM imply that significant political–institutional change is needed at the international level in order to prevent global environmental catastrophe. Conservative forms of EM theory envision both the extension of market regulation and the global convergence of all types of industrial activities on 'cleaner' (i.e. leading western) production models. Leaving aside the cultural chauvinism and political naivety of these assumptions, a clear consequence of the EM perspective is the need for global institutional mechanisms, such as environmental conventions and perhaps even an international environmental protection agency, which could condition the systems of capitalist production in ecologically sustainable ways. Stronger versions of EM theory, such as Beck's risk society thesis, argue the need for an explicit democratisation of industrialism and technological change. Beck is well aware of the ability of capital to use uneven ecological regulation to its advantage and his analysis certainly aspires to the democratisation of scientific and productive systems at all levels – including the global – through the proliferation of critically aware political fora.

Ecosocialism

A third contemporary environmental institutional disourse is the rather loosely defined ecosocialist perspective – a diverse set of analyses and prescriptions which none the less can be clearly distinguished from ME, EM, and other political ecologies, such as ecocentrism and ecoanarchism. Ecosocialists include a broad grouping of Marxian theorists in western (e.g. Altvater, 1993; Frankel, 1987; J. O'Connor, 1994; M. O'Connor, 1994; Peet, 1991; Pepper, 1993), and developing countries (e.g. Mies and Shiva, 1993; Shiva, 1991; Sundararajan, 1996). The discourse has two defining features: first, its analytical premises are rooted in Marxian political economy (though a considerable interpretative range is evident, ranging from near-orthodox historical materialism to liberalised variants, such as critical theory); and second, from this, the capitalist mode of production is seen as the main, if not the exclusive, source of recent and contemporary ecological crises. Given the profundity of these criticisms of market society, ecosocialist theory is at least by implication, if not always by prescription, a transformative perspective that condemns capitalism from a 'radical, socially just, environmentally-benign – but fundamentally anthropocentric – perspective' (Pepper, 1993: xi).

The ecosocialist position is neatly encapsulated in the statement by Pepper (1993: xi–xii) that 'Social justice . . . or the increasingly global lack of it, is the most pressing of all environmental problems'. As will be seen, this prioritisation of justice over other ethical concerns, both as an analytical focus (i.e. injustice is the chief source of ecological crisis) and a political goal, sets ecosocialists at odds both with conventional environmentalism (i.e. ME and EM) and ecocentrism. The ecosocialist perspective was prefigured in a seminal essay by Stretton (1976) which, while recognising the seriousness of ecological problems, argued that class and social justice must be pivotal political–conceptual axes of environmentalism. Stretton's essay sounded the environmental tocsin for the western left, arguing that the socialist movement must embrace the ecological question as a key consideration in social wellbeing. However, Stretton cautioned the left against absorption within the mainstream environmental movement, urging only a 'rational alliance' (1976: 13) between socialists and Greens. For Stretton and many subsequent socialist ecologists (e.g. Bahro, 1982; Weston, 1986; Harvey, 1993a,b), the fact that the environmental movement was, and often remains, detached from progressive class politics means that green ideals may conflict with social justice. Thus, socialist analysts have attacked many green policies for their socially regressive implications, reflecting the suspicion of many in the left political tradition – especially within organised labour – that environmentalism too closely reflects the interests of the bourgeoisie, the class which has dominated many of the new social movements (Dobson, 1990).

Pepper continues this tradition of criticism, arguing that the anarchistic tendencies of ecocentric Greens distances them from the socialist tradition. In particular, Pepper insists that Greens cannot join the ecosocialist project until they eschew 'those aspects of their anarchism that are more akin to liberal and postmodern politics' (1993: 3). At the same time, however, Pepper acknowledges that Marxism must accommodate certain green concerns that are in fact immanent to the socialist tradition itself, 'including traditions of decentralism and of the society–nature dialectic' (ibid.). Pepper thinks that the more progressive elements of anarchism might thus provide a useful corrective against the tendencies to totalitarianism evident in certain socialisms (see, for example, Bookchin, 1995a,b).

Just what, then, is the 'ecosocialist project'? As yet, this is a difficult question to answer given the inchoate nature of ecosocialist theory and politics. Indeed, Pepper suggests that 'a major and urgent task' still confronting ecosocialists is the enunciation of the '*details* of a green socialist political economy' (1993: xii, his emphasis). None the less, he argues, ecosocialism already has more to offer ecological politics 'than just an incisive analysis of capitalism' (ibid.: 3). The ecosocialist analysis includes:

> a dialectical view of the society–nature relationship, which is not like that of ecocentrics or technocentrics, and challenges both of them. It has a historical materialist approach to social change which ought to inform green strategy. And it is committed to socialism.
>
> (Pepper, 1993: 3)

Moreover, as Merchant (1992) points out, ecosocialism offers a clear, 'homocentric' alternative to ecocentric analyses of environmental crisis. As Pepper puts it, ecosocialism: 'is . . . anthropocentric enough to insist that nature's rights (biological egalitarianism) are meaningless without human rights (socialism). Eco-socialism says that we should proceed to ecology from social justice and not the other way around' (1993: 3).

Ecosocialism has thus far produced a considerable body of analytical work that has addressed the causes of ecological crisis in capitalist and (former) socialist societies. Unlike ME and EM, ecosocialist analysis embraces the complexities of the international political economy, focusing on the ecological and social consequences of the uneven process of economic globalisation. Thus, ecosocialists, and other radical, political ecologists, have opposed conventional development theory by pointing out that underdevelopment, including poverty and environmental degradation, is a necessary feature of international capitalism. Thus, rather than alleviating regional underdevelopment as its proponents claim, global free trade is explained by ecosocialists as a way of transferring wealth – including ecological assets – from peripheries to core nations. Moreover, recalling the analysis of Chapter 5, Altvater (1993; 1994) has highlighted the reverse flows in this unequal exchange of environmental quality through trade which transfers waste from core to peripheral nations, from rich to impoverished regions.

Recent collections of ecosocialist thought have addressed the question: is capitalism sustainable in any meaningful sense? (See M. O'Connor 1994; and the 'Marxism and Ecology' issue of *Science and Society*, pp. 60–3 (1996).) In general, the answer has been no, although some analysts elsewhere, such as Sandler (1994), have conceded the likely emergence of a 'green capitalism', a nightmarish vision of a global market society that combines ecological equilibrium with social misery and inequality. However, in response to Sandler's distopian vision, Schwartzman (1996: 261–2) doubts that 'full ecological sustainability could coexist with mass poverty . . . since the two are strongly incompatible at present'.

In general, ecosocialists identify two ecologically destructive tendencies that are endemic to capitalism (not all theorists stress both causal mechanisms, or emphasise them equally). The first anti-ecological tendency is sourced in the economic immiseration and inequality caused by capitalist social relations of production – social and ecological exploitation are seen as intimately related. As Weston (cited in Dobson, 1990: 171) states, the cause of 'virtually all environmental problems, both physical and social, is poverty'. According to Dobson:

> Many socialists will analyse phenomena like deforestation from just this point of view – the fundamental problem is much more one of inequitable land distribution (which produces slash-and-burn farmers) and structural poverty (which produces periodic but highly damaging jungle gold rushes), than it is one of an insatiable and environmentally-insensitive desire to eat hamburgers.
>
> (Dobson, 1990: 172)

However, as Dobson notes, many Greens and Marxists doubt that socioeconomic redistribution will prevent environmental degradation. It *is* possible to imagine an egalitarian world that pursued a limitless exploitation of nature (but by definition this would probably not be a *capitalist* world).

The second anti-ecological tendency of capitalism identified by ecosocialists emerges from the productive logic of market society itself. J. O'Connor (1994), for example, emphasises the contradiction between the forces/relations of production and the material 'conditions of production' as a key source of environmental degradation in capitalism. O'Connor's analysis echoes Marx's own observation that, 'Capitalist production . . . develops technology, and the combination together of various processes into a social whole, only by sapping the original sources of all wealth – the soil and the labourer' (cited in Parsons, 1977: 174–5).

Altvater (1993; 1994) identifies the accumulative logic of capitalist valorisation as a source of ecological exhaustion and ruin. Altvater has developed a scientific analysis of the exhaustive, and ultimately destructive, exchange between self-expanding markets and their ecological conditions of existence. Altvater and other ecosocialists, such as Lovejoy (1996) and Schwartzman (1996), employ the scientific theory of thermodynamics to demonstrate that capitalist self-valorisation, both concretely and ideologically, takes no account of the inescapable tendency for entropy (energy loss and corruption) in the material world. Altvater argues that markets mobilise natural inputs over a comparatively short time and thereafter these resources 'are available only in quantitatively reduced and qualitatively degraded forms – or, indeed, they are completely and irreversibly "used up", consumed' (1994: 85).

Indeed, the law of value actually demands an ever-increasing rate of entropy, revealed materially as worsening resource depletion and an expanding quantum of waste that is either too hazardous or too corrupted to have use value (toxic waste is an 'anti-use value', in that its 'use' will diminish the value of any user's existence). Of course, entropy is an inescapable fact for any human society. Altvater's point is that the logic of accumulation is the antithesis of the 'system intelligence' that is needed to constrain the rate of energy loss which arises from human social activity. If globalisation has proved that 'capitalist growth and spatial expansion has no inherent borders', the fact of increasing ecological ruin demonstrates that capitalism is 'limited by external factors', principally, the process of entropy (1994: 88).

Altvater bitterly opposes neo-classical economic theory (and its variants, such as ME) 'as the explanatory discourse for the market system'. Neo-classical theory is both aspatial and timeless – facts which prevent this framework from comprehending the ecological basis of human social activity. Moreover, he argues that conventional economics actually assumes, and therefore encourages, the increase of entropy (scarcity) in social systems – the discourse of economics is therefore deeply anti-ecological.

As Marx noted, uncosted, or common, resources are 'a free gift of Nature to capital' (cited in Parsons, 1977: 171). However ecosocialists, such as Altvater, argue that to assign prices to uncosted or common aspects of nature – the ME way of 'valuing the gift' – is to draw these resources even further into the ecologically destructive market system, not to prevent their exhaustion. Pricing merely assigns to nature exchange values, not use values, and thus serves only to rationalise resource allocations socially, not ecologically. Thus, the absurdity of pricing as a solution to environmental exhaustion and pollution stems from: 'the radical gulf between the ecological logic of (irreversible) transformations and the economic logic of commutativity, equivalence in exchange, and equilibrium' (Altvater, 1994: 90n).

Altvater's astute assessment is that ME fails really to solve the ecological problems of uncosted resources. The effect of market valuation is rather 'to conjure them away' (*ibid.*: 86). Moreover, this theoretical conjuring contributes to a broader ideological and material project through which all existence is commodified, and humanity thus alienated both from itself and its natural basis. As Engels observed of this process a century ago: 'To make the earth an object of huckstering – the earth which is our one and all, the first condition of our existence – was the last step toward making oneself an object of huckstering' (cited in Parsons, 1977: 172).

If the future is threatened, what can be done to restore social and ecological security to an increasingly degraded and unequal globe? By condemning capitalism as fundamentally anti-ecological, ecosocialists suggest – though sometimes only by intimation – that a radical transformation of political–economic structures is needed. In short, a new, post-capitalist social formation is required.

Several visions of ecological socialism have been forwarded (e.g. Frankel, 1987; Pepper, 1993; Ryle, 1988), but these appear both schematic and politically distant, not yet supported by mass politics. The ecosocialist vision of this new alternative social formation remains clouded by analytical uncertainties and the succession of political defeats which the Left has endured in many parts of the globe in recent years. Note the hesitancy in Schwartzman's assessment of the way forward:

> The possibility of a sustainable future that includes the goal of meeting global human needs is of course contingent on viable strategies to radically constrain capital, perhaps to the point of constituting the transitional society defined as 'socialism' in classical Marxism'.
>
> (Schwartzman, 1996: 264)

The gloomy equivocation of ecosocialists is understandable if one considers the enormous political gulf between the present 'triumph of the market' (Altvater, 1993) and the ideal of an 'ecological socialism' (J. O'Connor, 1994). Some, such as Sandler (1994), wearily predict the triumph of ecological modernisation and the emergence of a new 'green capitalism' that will intensify social division and only delay the day of environmental reckoning. Others, such as Schwartzman, find the reformist path tempting, if not inevitable: 'Perhaps part of this debate is really semantic: at what point does a society with progressively constrained capital become worthy of being called "socialist"? When does radical reform become revolutionary . . .' (Schwartzman, 1996: 264.).

Schwartzman realises that the structural presence of capital, however 'constrained', excludes the possibility of socialism (this is suggested in an acknowledgement that his query is a *deja vu* of an old debate' (Schwartzman 1996: 264)). Altvater's own work reflects a similar asymmetry between transformative analysis and reformist-leaning politics. He implies that reformist 'conditioning' of market processes may be necessary both to prevent imminent ecological catastrophe and to establish the political possibilities for the eventual transformation to ecological socialism:

> We must create social and political border lines before the frontier of capitalist expansion reaches the last ecological border, which would be

173

fatal to the conditions of the survival of the human race. Once we realize that a transformation in the *social* forms is what is required, fruitful discussions about ecological reform might begin . . . .

(Altvater, 1994: 89, original emphasis)

In summary, ecosocialism envisages profound changes to both the political–economic and political–institutional frameworks of globalising capitalism. At the very least – i.e. in its most reformist guise – ecosocialism echoes Beck's call for the 'political dethronement' of science and technology, while also insisting upon the democratisation of markets and productive systems. The political consequences of this view are stronger ecological regulatory mechanisms at at all political scales, a redistribution of wealth from rich to poor nations, radical constraints on the autonomy of transnational capital and democratic control of productive and technological change. These consequences all point toward the establishment of strong and democratic global institutions – perhaps beginning with a re-chartered United Nations – which would be guided by the ideal of environmental justice; viz., a society of societies which could guarantee that its variously constituted citizens enjoy environmental wellbeing, socio-economic security and freedom from cultural or political oppression. Importantly, such a supra-society would be both prepared and empowered to act against environmental injustice.

Ecosocialists have emphasised the antipathy between a self-picture that relates the person primarily to other humans and one which merges the self with all of nature. This antipathy follows rather strongly from Pepper's work, which thus emphasises the potential contradiction between environmental and ecological justice. Benton, on the other hand, argues in less absolute terms for a prioritisation of human interests within a continuum of moral considerability. This may not appeal to those who insist on strict bio-egalitarianism but, as argued in Chapter 6, such a moral continuum has a practical application, and can form the basis for reconciliation between green and socialist positions. It also appears to be consistent with the view of the human–nature relation advanced by Marx and Engels.

The suggestion that radical transition must first await the remedial and mediating work of reform is hardly new to progessive politics – but it may have practical merit in the present conjuncture. Institutional reform of a fundamental kind is implied not only by ecosocialism, but also by the deeper kinds of EM analysis. Further, if the pursuit of the ME agenda is to involve a radical restructuring of national and international institutions worldwide to increase the scope of market transactions, then this move, too, would seem to require public debate in a democratic global forum. If so, what institutional and political forms should such 'market conditioning' take? What should be the 'next steps' in regulating global capitalism?

THE INTERNATIONAL POLITICAL ECONOMY AS A SYSTEM OF GOVERNANCE

We are governed today not only by the constitutions, democratic or otherwise, of nation states but increasingly by an international political system. There is a worldwide *web* of governance. This web cannot be simply swept away, to be replaced by some political utopia, whether Green or anarchist or socialist. We have seen the dangers of revolutionary transformation by force in this century and, even if violent revolution were feasible at global level (which it is not), it cannot be considered desirable. The best solution must be one in which the dialectics of justice, discursive democracy in some form, can be brought to bear on the formation of institutional solutions. As we saw in the last chapter, the dialectic has already been transformed by an extended vision of the self such that the non-human world must now be included in considerations of justice viewed as a relationship between 'self' and 'other'.

However, just solutions to the kind of problems we face today – environmental and ecological – cannot be found by withdrawing inside the nation state or into any other localised community. As long as there is global capitalism – and a global market – there must be a countervailing power of similar scale to provide the aegis under which an environmentally and ecologically just society of societies or 'commune of communes' (in Bookchin's terms) may gradually take shape. What institutional form this global society will take cannot be predicted. That will depend on the progress of the dialectic. Perhaps nation states will continue to have a role. Perhaps, as seems possible today, the primary unit of governance, and thus democracy, will be of much smaller scale, cities or bioregions (see, for example, Bookchin, 1995a; or Sale, 1985). Perhaps new transnational scales of governance will increasingly develop (such as the European Union). The present system is not a static one. It is constantly changing – albeit slowly. What we have to do is understand the nature of the international political system as a system of governance in *process of change*, and the role of nation states in that system. We can then define a 'next step' in a progressive evolutionary and incremental strategy headed *in the direction* of environmental and ecological justice; a direction in which human communities and individuals can *increasingly* flourish in harmonious relations with each other and with the non-human world.

International governance

In the Middle Ages the area of the world now called Europe was a mosaic of territories controlled by a variety of powers (princedoms, duchies, church estates, monasteries, cities, kingdoms) loosely held together under the theocratic rule of the Papacy and the Holy Roman Empire. 'Europe' as Held remarks, 'more accurately meant Christendom'. Feudal aristocrats with military power ruled over territory in the countryside, quasi-independent governments of cities controlled commerce and industry under their own charters. The political theory and political legitimacy of all authorities was derived from Christian doctrine. 'The Christian world-view transformed the rationale of political action from an earthly to a theological framework; it insisted that the good lay in submission to God's will' (Held, 1995b: 33). Gradually the powers of local rulers became concentrated in the hands of monarchs ruling uniformly over larger territories. Between 1500 and 1900 the number of separate and independent political units in Europe dwindled from about five hundred to twenty-five (*ibid*.: 32; Tilly, 1975). The 'sovereign' state, ruled either by absolutist or constitutional monarchs emerged:

> Absolutism signalled the emergence of a form of state based upon the absorbtion of smaller and weaker political units into larger and stronger political structures; a strengthened ability to rule over a unified territorial area; a tightened system of law and order enforced throughout a territory; the application of a 'more unitary, continuous, calculable and effective' rule by a single, sovereign head; and the development of a relatively small number of states engaged in an 'open-ended, competitive, and risk-laden power struggle'.
>
> (Held, 1995b: 35, citing Poggi, 1978: 60)

The struggle within the national territory was between absolute monarchy and the variety of powers which were strengthened by production and trade. These powers, as Mann (1986) argues, were not just the citizens of towns, the burgers or bourgeoisie, but a nexus of interests in private property and production, from the peasant to the small aristocrat. Capitalist relations under norms of property rights and trading contracts developed first in agriculture. Cities merely provided the essential nodes in a network of trade. By the end of the fourteenth century, Mann writes: 'individual families and local village-and-manor communities were participating in a wider network of economic interaction under institutionalised norms governing property possession, production relations and market exchange' (1986: 409). Medieval dynamism, he writes, 'took the form of a drive towards capitalist development' (*ibid*.: 412).

Here, then, we have the makings not only of the sovereign nation state but of an international system. Absolute monarchy brought with it the development of an administrative apparatus, the beginnings of professional bureaucracies, and standing armies. Wars between states increased the salience of territorial boundaries marking off friendly from enemy territory. Political identity marked by territory and political borders grew with conflicts between states (see Tilly, 1975: 73–4). Preparation for war was the great state-building activity. The Thirty Years War, which wrought the most appalling destruction in Europe, combined the struggle between competing dogmas with the struggle between competing national and imperial forces.

The Peace of Westphalia which brought to an end the German phase of the Thirty Years War is usually regarded as a landmark in the development of the international system. Held (1995a) regards the 'Westphalian' model of international order as lasting from 1648 (The date of the Peace) to 1945 with the end of the Second World War and the founding of the United Nations. Westphalian principles, however, remained embedded in the succeeding United Nations model. Held summarises the principles of the Westphalian model as follows:

> It depicts the development of a world community consisting of sovereign states which settle their differences privately and often by force; which engage in diplomatic relations but otherwise demonstrate minimal cooperation, which seek to place their own national interest above all others; and which accept the logic of the 'principle of effectiveness', that is, the principle that might eventually makes right in the international world – appropriation becomes legitimation.
>
> (Held, 1995a: 104, citing Cassese, 1986)

The Westphalian order presupposed that domestic policy could to a considerable degree be protected from the influence of international events, thus providing for the nation state a domain of sovereignty and autonomy over its internal affairs. Held (1995b) distinguishes between sovereignty and autonomy, and between internal and external sovereignty. *Internal* sovereignty means the right of a national government to 'enjoy the final and absolute authority within a terrain'. External sovereignty means that there exists no final and absolute authority above and beyond the nation state. The assumption of external sovereignty establishes the rationale for war where states cannot reach agreement to resolve differences by negotiation. It is equally the basis for a negotiated order, since an international regime governing some aspect of international relations can only be established by consent of the states.

Autonomy does not refer to the 'right' but to the actual capacity of a state to implement policy: 'the capacity of state managers and agencies to pursue their policy preferences without resort to forms of international collaboration and cooperation' (*ibid.*: 100, see also Held and McGrew, 1993; Goldblatt *et al.*, forthcoming). A state may have the right of sovereignty but little actual capacity to realise its sovereignty through its policies.

The principles of sovereignty and autonomy did not apply between the colonial powers and their colonies in the various imperial systems which developed from the seventeenth century onwards. But at least it applied approximately to the nations of Europe. However, the global disaster of the

Second World War and the development of weapons of mass destruction made a change in the Westphalian system most urgent. The system was substantially modified by the introduction of the United Nations as a global governing body, together with an agglomeration of international institutions focused upon particular issues under the aegis of the UN. The United Nations system marks a transition from a world of national sovereign states, settling their differences by force, to one marked increasingly by international and in some instances negotiated global regimes established for specific purposes with the consent of nations.

The United Nations system was designed both to recognise the reality of the distribution of power among states at the end of the War, and to encourage existing nations to settle their differences without resort to war, thus by negotiation. Moreover, the entry of many new actors into UN arenas brought on an agenda of decolonisation which spread the principle of national sovereignty much more widely. The collapse of the Soviet imperium further reduced the number of client states in the world order, and by the 1990s very little remained of the European empires formed around core nation states.

What we have today, then, is an international system of governance in which, putatively, sovereign nation states determine the regulation of capitalist enterprise within their borders and negotiate freely, and as equals, among themselves to establish international regimes for the regulation of development which has an impact beyond their borders. The key questions we have to ask, then, are whether this negotiated order is adequate to advance the ends of environmental and ecological justice, and, if not, what further institutional change is needed?

The negotiated order

An extensive network of treaties, conventions and other instruments and principles of international law now exists to regulate environmental exploitation. International regulatory institutions and bargaining over environmental regulation have been the subject of much recent study (for example, the collection edited by Hurrell and Kingsbury, 1992; Haas *et al.*, 1993; Sjöstedt, 1993; Young, 1994). New institutions are constantly being created and new principles (e.g. the precautionary principle, intergenerational rights, environmental crime) are beginning to penetrate the arena of negotiation. Birnie (1992: 83) considers that 'using the existing sources and concepts of international law and the wide range of concerned organisations, an identifiable environment-specific regulatory regime has emerged that can be regarded as "International Environmental Law"'. Birnie concludes, however, that states are not yet ready to abandon the principle of sovereignty in the interests of what we have termed environmental and ecological justice, although they are prepared to allow limitations upon sovereignty (a concept of 'reasonable sovereignty') necessary to secure environmental goals.

As to international bargaining and the institutional mechanisms of regulation that result, research has been focused on the 'effectiveness' of international regimes. Keohane *et al.* conclude that, 'International institutions do not supersede or overshadow states. They lack resources to enforce their edicts. To be effective, they must create networks over, around, and within states that generate the means and incentives for effective cooperation among those states' (1993: 24). These authors conclude that environmental issues have been considered separately, 'independently of possible underlying causes such as population growth, patterns of consumer demand, and practices of modern industrial production' (*ibid.*: 423). They point out that there has been no attempt to reform the global political system to ensure that the costs of transboundary pollution, for example, are incurred by its originators; and that although there has been a considerable degree of success and co-operation in some areas of environmental disorder, in others there is a story of continual failure. Control of international oil pollution of the oceans has been ineffective, the regulation of fisheries has not prevented overfishing, and, significantly, 'until very recently, the World Bank has resisted environmentalists' attempts to make it "green" ' (*ibid.*: 424).

In Young's analysis, environmental regimes are based upon contracts into which multiple actors voluntarily enter in order to govern certain activities under their control (Young, 1994). Uncertainties inevitably attend the negotiations. There is initial uncertainty as to the identity of the participants. In addition there is likely to be uncertainty about the alternative strategies available to the participants and about the outcome associated with various combinations of choices. Therefore there are some definite incentives for the participants to try to identify 'mutually beneficial deals' (win–win solutions). Young cautions that this does not eliminate the distributive aspects of bargaining or the role of power in efforts to achieve distributive advantages. But it does provide a 'counterweight' in favour of 'integrative bargaining' (*ibid.*: 101). There are also uncertainties stemming from the fact that it is regulatory *institutions* – sets of rules governing a wide range of situations – that are being negotiated rather than specific deals. Uncertainty, he points out, 'has the effect of increasing interest in the formation of arrangements that can be justified on the grounds that they are fair in procedural terms whatever substantive outcome they produce' (*ibid.*: 43).

In environmental bargaining, then, we have examples of the application of Rawlsian principles. There is something like an initial position in which the participants are approximately equal and have to make decisions about a constitution whose outcome for them is 'veiled'. Admittedly this 'veil of uncertainty' is less opaque than Rawls's 'veil of ignorance'. But there appears to be empirical evidence that, even in situations of 'uncertainty', actors adopt a strategy of minimising risk.

As to how successful institutional bargaining can be in achieving its own limited objectives, Young is cautious: 'Like self-interested actors in all social arenas, those attempting to work out the terms of international regimes are often stymied by impediments to bargaining that can prolong negotiations over institutional arrangements and that can easily end in deadlock' (*ibid.*: 106). He defends six 'hypotheses' about factors which increase the likelihood of success. Success is more likely when: (i) the issues at stake lend themselves to treatment in a contractarian mode; (ii) arrangements are available that all participants can accept as equitable; (iii) solutions are describable in simple terms; (iv) clear-cut and reliable compliance mechanisms are available; (v) exogenous shocks or crises occur and are perceived; (vi) effective entrepreneurial leadership by individuals is available. Unfortunately these are strictly contingent variables. We cannot count on exogenous shocks coming along in time to rectify the cause of the shock. We cannot (*pace* Beck, 1992) count on the existence of uncertainty to produce justice. The more that is known about an issue such as global warming, the more will its distributional effects become apparent.

Empirical studies of international regimes will continue to be important but this sort of work cannot by itself answer the wider question of the justice of the existing international framework of governance. How much that is important for environmental and ecological justice, for example, is *not* covered by *ad hoc* regimes? We discussed in Chapter 1 significant examples which were not adequately covered by the negotiated order. Moreover, as Young is ready to admit (1994: 160), judgements about regime effectiveness depend on larger world-views (i.e. of justice), which analysts bring to the study of the subject. What we have to consider is the relationship between instrumental effectiveness and the wider context of governance. This context contains a mixture of institutions and supportive ideas about the role of *government*, and not just governance, in the achievement of justice.

Shue (1992) argues that environmental justice will have to become a major consideration of future negotiations over climate change. This is because, despite our best efforts to prevent it, global warming is going to occur. There is little doubt that global warming will have sharply varying local effects (see Intergovernmental Panel on Climate Change (IPCC), 1996). Shue argues that, in the interests of justice, there must be international provision to help the most affected nations 'cope' with the effects of global warming. Shue asks: 'What are the individual interests that the poorest nations would be asked to sacrifice if they were asked to ignore provisions for coping? They are, in a word, vital interests, survival interests' (1992: 394). He posits that we cannot in justice ask nations to sacrifice vital inter-ests in order to maintain interests in the richer nations that are not only *not* vital but trivial. If justice is to be done, there will have to be a global transfer of resources from the rich to the poor nations to help them cope with the problems of global warming. Under present institutional condi-tions, Shue considers only a minimal 'guideline' practicable. This guideline is that, 'poor nations should not be asked to sacrifice in any way the pace or extent of their own economic development in order to help prevent the climate changes set in motion by the process of industrialisation that has enriched others'. Shue does not mean to ignore environmental damage from development, but if development is not to be prevented for reasons of ecological justice (e.g. the halting of destruction of the Brazilian rainforests, the limitation of coal-burning power generation in China or the prevention of mining in Papua New Guinea), then the poorer nations concerned must be compensated by the richer.

If compensation is to be provided, there must be a cash flow. Various methods of taxation have been suggested, for example, the proposal by Nobel-laureate economist James Tobin for a 0.5 per cent tax on international flows of money which would raise more than $1.5 trillion annually (see French, 1995: 185, citing Walker, 1993; Childers and Urquhart, 1994). But if there is to be taxation, as the old American slogan goes, there must also be representation. Taxation requires the authority of a world body equipped with much more legitimacy and popular support than the UN can at present muster.

While it can hardly be doubted that an extensive 'new world environmental order' (in the words of Levy *et al.*, 1993: 425), has developed incrementally over the last fifty years, this order is fragmented and fragile and is increasingly contradicted by the much more centralised order of world economic governance. There is no world government but there *is* world governance. The global economy is governed by a nexus of institutions such as NAFTA, The World Trade Organization (formerly GATT), The World Bank, the IMF. The step of separating 'the economy' from politics itself has major ethical implications – as though somehow our material wellbeing and the wellbeing of our environment has nothing to do with what we debate about and struggle for politically. 'The economy' is seen in the very narrow terms of entitlement theory and utilitarianism. Wellbeing is equated with commodity production and exchange, and the level of business activity measured by GDP. But this separation is quite artificial – little more than an ideological smokescreen designed to conceal the real injustice of the distributive principle – 'adiaphorization' in Bauman's language. The decisions and rules created by the economic order affect every aspect of the lives of people, and every aspect of the global environment .

There can be little doubt that the struggle is on for control of the agenda of world governance. Nader points out that 'it is only recently that corporations developed the notion of using trade agreements to establish autocratic governance over many modestly democratic countries' (1993: 2). From 1986 and the Uruguay Round of trade negotiations, Nader claims, 'multinational corporations thrust an expanded set of concerns on GATT that went far beyond traditional trade matters' (*ibid.*). The struggle for control is conducted largely in secrecy and without input from citizens or non-government organisations. The first element of the agenda is now in place:

the World Trade Organization. The WTO has already enacted rules which will have a devastating impact on the environment. Under these rules national governments are not permitted to restrict imports which have been produced under environmentally damaging conditions. Hence the USA was prevented from banning importation of tuna caught with methods which kill large numbers of dolphin (as mentioned in Chapter 5). This also means that, when subjected to national regulations controlling pollution at the source of production, multinational corporations can avoid upgrading their technology and simply shift the polluting plant to a country which accepts lower standards. Mexico is a case in point where production processes have turned industrial areas into cesspools of disease and heavy metal pollution (see Watkins, 1996). Of course, if imports produced cheaply are allowed to compete with locally produced products subject to pollution controls (as well as all sorts of other ethical standards), this puts immense pressure on national governments to lower their own standards. The result will be a steady decline to a lowest common denominator of environmental regulation, the exact opposite of that to which the nations of the world put their signature in Rio de Janeiro in 1992.

Yet it must also be recognised that world trade is, however unevenly, spreading more widely the kind of living standards people in western, now post-industrial, nations have enjoyed for decades. Paradoxically though, these standards were achieved, not by free market capitalism, but by the state forcibly transferring capital from private profits to public purposes and – under pressure from the working class via representative democracy – guaranteeing reasonable wages and decent working and living conditions. Along with a vast array of electronic and mechanical gadgetry, 'western' living standards also include essentials like better public health, housing, education, political participation, income security, opportunities for creativity and a wide choice of lifestyle. Some people in the west may be beginning to find the normal range of possible 'western' lifestyles in the end unsatisfying, that the reduction of 'self' entailed is demeaning. But then they still have the choice of living otherwise. For how much longer these standards will be maintained for the majority is very much an open question. For some in 'developed' nations they have already been lost. The institutional conditions under which welfare states were created to cater for basic needs have now largely disappeared.

The lifestyles of poorer countries may be more ecologically sustainable than those of the developed world. The living standards of the rich countries may be far from ecologically just. But we cannot ignore the fact that the latter are the object of desire of many in the Third World. Certainly world trade under a global capitalist regime is creating class societies, sometimes with immense polarisation between rich and poor within the nation. Even internationally there are huge differences between the potential for prosperity of some of the new industrial countries, such as those of East Asia, and those nations excluded from the global economy, such as many of the African nations. Under the present system of global governance without government, there is no hope of resolving the enormous contradictions and ethical dilemmas that are thrown up by the uneven development on which capitalism thrives. In truth the harshest contradiction is perhaps between the demands of environmental justice – the need to spread good environmental conditions, and those of ecological justice – the protection of the planet for the flourishing of all life forms and ecological systems. These contradictions will grow more dangerous and threatening in the next century and increasingly in need of resolution if ecological destruction and perhaps devastating local wars are to be avoided. So we must consider wider global institutional change.

Cosmopolitan democracy

Held has observed a disjuncture between sovereignty and autonomy. To take one of the examples considered in Chapter 1, it is evident that, in nuclear testing, France and other nations (e.g. China) have actively asserted their sovereignty above the common heritage in the planetary environment. Therefore what seems to be required, if nuclear testing is to be prohibited, is a further reduction in sovereignty and the institution of an international regime under the UN framework to end such testing. The basis for such a regime already exists in the form of test-ban treaties. A comprehensive regime banning nuclear testing in the atmosphere, sea and space has been in process of negotiation for thirty or so years. It took a further step forward in July 1996. The only respectable rationale for the armaments industry is the Westphalian principle of effective power and the competitive order of sovereign states. Yet the continued existence of the armaments industry brings into focus the combination of national *sovereignty* with diminished national *autonomy*. In nuclear testing, France is asserting its national sovereignty but also acknowledging its economic dependence upon the armaments industry, a form of 'nautonomy' to use Held's term.

Advocates of cosmopolitan democracy argue that the negotiated order rests on assumptions that are unrealistic in today's world. Held (1995b) argues that five 'disjunctures' have developed in the last fifty years between the idea of the sovereign state as in principle capable of determining its own future, and the realities of international power. These disjunctures are to be found in the development of international law, the internationalisation of political decision making, the existence of hegemonic powers and international security systems, the globalisation of culture and the increasing integration of major elements of the world economy.

The point which Held wishes to establish is not that states are unequal in power, which is trivially true, but that they come to negotiate in the presence of a nexus of pre-existing international ties: 'The operation of states in an ever more complex international system both limits their autonomy (in some spheres radically) and impinges increasingly upon their sovereignty' (*ibid.*: 135). In effect, Held's argument is that there already exists a form of international governance not unlike that of pre-modern Europe. The form of international governance is not a centralised, pyramidal 'state' based on the model of monarchy. International governance consists of a many-dimensioned nexus of authority and power. The international 'state' does not rule over a clearly defined territory. Different territories are subject to different aspects of its rule. There is, in short, a mosaic of territories controlled by a variety of powers.

Held's critique proceeds like that of Rawls from the principle of autonomy via a 'democratic thought experiment'. This experiment conceives of the constitutional rules a group of persons would agree upon to govern their relationships under conditions in which they are free 'in equal measure', that is to say where coercive relations of any sort are wholly absent. The conditions of such an experiment are those which will be familiar from similar experiments postulated by Rawls (1971; 1993a) and Habermas (1990).

The democratic thought experiment is used to justify the principle of autonomy. Held writes that the principle can be stated thus:

> Persons should enjoy equal rights and, accordingly, equal obligations in the specification of the political framework which generates and limits the opportunities available to them; that is, they should be free and equal in the determination of the conditions of their own lives, so long as they do not deploy this framework to negate the rights of others.
>
> (Held, 1995b: 147)

This is, of course, a principle of personal autonomy. This principle is the foundation of democracy. The invention of democratic institutions results from the working through of this principle. Conversely, the justification for democratic institutions can be sourced to this principle.

Because the autonomy of the nation state is restricted by the web of international ties within which states are today enmeshed, the pursuit of personal autonomy cannot be conducted within the nation state alone. The constitution of nation states can no longer be made to respond to demands (for justice) arising from the principle. No-one wants an armaments industry whose purpose is to kill and maim people. But trapped within the prisons of their nation states people are subject to the prisoner's dilemma. It is a double dilemma. Not only is there the dilemma arising from security, there is also the dilemma arising from dependence on the armaments branch of capital. Thus, a new constitutional order is required which will enshrine the principle of personal autonomy through new institutions of democracy. This new democratic order must be cosmopolitan, that is to say it must extend beyond national boundaries and ultimately involve the whole world. 'Cosmopolitan democracy' does not entail a single global structure of governance, but it does entail a mixture of institutions encompassing different territorial scales, some of which will be global. Held suggests five institutional initiatives:

1 A global parliament (with limited revenue raising capacity) connected to regions, nations and localities.
2 A new charter of rights and duties locked into different domains of political, social and economic power.
3 Separation of political and economic interests; public funding of deliberative assemblies and electoral processes.
4 An interconnected global legal system embracing elements of criminal and civil law with mechanisms of enforcement from the local to the global; the establishment of an International Criminal Court.
5 Permanent shift of a nation-state's coercive capability to regional and global institutions with the ultimate aim of demilitarisation and the transcendance of the war system.

(Held, 1995b)

A more detailed agenda for reform of the UN system is provided by Archibugi (1995). As Held recognises, such a reform can only be undertaken step by step, much as the European Union was, and is still being, constituted. What could happen gradually is the transformation of governance – secretive, undemocratic, largely unseen – into government.

TWO CRITIQUES OF A GLOBAL CONSTITUTION

Transforming world governance into world government is in many ways a frightening prospect. While there are different national systems of government there is at least the possibility of escape ('exit' to use the political science term) for people who do not wish to be governed according to a particular system. World government might finally close all the exits. Since government is an expression of culture, different national systems of government to some extent guarantee cultural diversity. World government might greatly reduce cultural diversity. The oppressive potential of the administrative state is well understood, and this potential is immensely magnified at global scale unless it can be contained by democratic politics. However, the difficulties of effective democratic control of a global state – already formidable at the scale of large nations – are multiplied at global scale.

As we saw in Chapter 4, Rawls (1993a) has argued for a 'political conception of justice'. In *The Law of the Peoples* he sets out the political conception of right and justice which he considers should apply between nations to 'the political society of well-ordered people' (Rawls, 1993b: 68). He wants us to acknowledge that liberal democracy is not the only 'reasonable' social form:

that there are other ways of securing social representation and fairness than through the adversarial party political system and other western institutional forms.

Rawls thinks that only a set of 'general liberal ideas' are necessary to a law of the peoples. Explicitly, this means the abandonment of three egalitarian features which are common to 'strong' liberal formulations of justice (e.g. his own 'justice as fairness'): the fair value of political liberties, fair equality of opportunity and the difference principle. What remains is 'the veil of ignorance'. But the veil of ignorance is supposed to remove the class and culture-specific assumptions people might bring to constitutional decision making. Under the veil of ignorance, people confront one another as human persons, not as embodiments of cultures. So, if the veil of ignorance is to remain it is hard to see why the other principles which are presumed to follow from that condition are to be abandoned. Alternatively if the other principles are abandoned, why not the veil of ignorance also?

For Rawls the law of the peoples 'covers only political values and not all life' (*ibid.*: 38). The global extension of justice to future generations, incapacitated citizens, social groups (e.g. families), and 'what is owed to animals and the rest of nature' (*ibid.*) is explicitly ruled out of the discussion, though Rawls recognises the importance of these issues. However at a critical point in his essay Rawls argues for resource stewardship as an essential duty of every reasonable nation:

> An important role of a people's government . . . is to . . . take responsibility for their territory and the size of their population as well as for maintaining its environmental integrity. . . . They are to recognise that they cannot make up for [any] irresponsibility in caring for their land and conserving their natural resources by conquest in war or by migrating into other peoples' territory without their consent.
>
> (*ibid.*: 47–8)

While this argument recognises people's responsibility for their relationship with nature, it does not show any understanding of why, as in the case of Papua New Guinea, a nation may not be able to afford responsible stewardship, at least while aspiring to the sort of life held out as desirable by global capitalism.

After Kant, Rawls fears the potential for despotism of any form of 'world government', but thinks that there will be a need for 'various forms of cooperative association among democratic peoples' in order to secure the common global good (*ibid.*: 46). These new co-operative associations (he cites the UN as just one example) would in some cases have the power to sanction, economically or even militarily, any state which violates basic human rights. This role may also extend to include providing assistance for impoverished countries. Rawls says that 'in all *reasonably developed liberal* societies a people's basic needs should be met' (*ibid.*: 47, emphasis added).

But this worrying specificity allows him to discount the needs of those in countries which are not 'reasonably developed and liberal'. This circumscription of the right to the fulfillment of needs reflects a failure to grasp the political–economic origins of underdevelopment and dependency. In his later work Rawls further explains his position. When discussing 'background justice' he says:

> At some level there must exist a closed background system, and it is this subject for which we want a theory. We are better prepared to take up this problem for a society (illustrated by nations) conceived as a more or less self-sufficient scheme of social co-operation and as possessing a more or less complete culture. If we are successful in the case of a society, we can try to extend and to adjust our initial theory as further inquiry requires.
>
> (Rawls, 1993b: 272 fn)

Our view is that there is no such thing today as a self-sufficient nation with an enclosed and complete culture.

James Tully (1995) in a critique seemingly directed against the imposition of any form of uniform world constitution shows how the constitutional order of modern capitalism was imposed upon nations colonised by European powers in the eighteenth century. Tully (*ibid.*, ch. 3) identifies seven features of modern constitutionalism which support an 'empire of uniformity'. A constitutional state possesses an individual identity as a nation, an 'imaginary community to which all nationals belong and in which they enjoy equal dignity as citizens' (*ibid.*: 68). The 'corporate identity of nation and nationals in a state is necessary to the unity of a modern constitutional association'. Sovereignty entailing equality applies to citizens before the law within the nation state and equality among nation states. A 'modern constitution' thus comes into being at some founding moment and stands behind – and provides the rules for – democratic politics and economic competition.

In the uniform constitution, principles of justice are fused with ideas about what constitutes 'modern society' and the superiority of the European conception of modernity in particular. A facade of cultural uniformity, Tully argues, was created with the help of abstract models of political justice together with foundationalist and progressivist ideas of modernity to justify eliminating the multiplicity of political cultures and conceptions of justice encountered in the course of European expansion. His argument is supported with well-documented examples from North America showing precisely how the arguments of constitutionalists such as John Locke were deployed to justify dispossessing the indigenous peoples of their land and culture.

From the cultural traditions marginalised by the uniform constitutional order and embodied in the nascent negotiated order, Tully recovers three principles: mutual recognition, consent and continuity. The step of recognising the cultural tradition of 'the other' as valid in any negotiation appears straightforward, but it is perhaps the most difficult step. It is difficult because the process of 'understanding' is unavoidably shaped by the language of one's own cultural tradition. One cannot just step outside this language because there is no 'outside'. What mutual recognition seems to demand is first a recognition of one's own political cultural language *as* a cultural tradition and not a set of universal truths which transcend culture. Second, recognition means finding a place for the traditions of the other within the world created by one's own political language.

The principle of consent is among the oldest European principles of constitutionalism and legitimacy – as opposed to mere conquest and the imposition of rule. The maxim *quod omnes tangit ab omnibus comprobetur* (q.o.t.) – what touches all should be agreed to by all – was established long before it was taken up and incorporated systematically in liberal political beliefs. Tully observes that 'the form of consent should always be tailored to the form of mutual recognition of the people involved' (*ibid.*: 123). He cites examples of negotiations in encounters between Europeans and indigenous Americans in which the principles of mutual recognition, consent and continuity have been put into practice (e.g. the 'Two Row Wampun Treaty' between the Haudenosaunee Confederation and the Canadian Government in 1983).

Today mutual recognition means respecting the cultural roots of authoritarian regimes and frequently tolerating, without condoning, authoritarianism. The uniform constitutional order of international capitalism does not oppose such regimes. This should not surprise us. The government of the European colonies was completely lacking in democracy. Far from being indigenous, many authoritarian regimes today are the product of colonial rule which was the only model of effective government available when the colonial yoke was thrown off. Certainly the post-war history of independent Singapore strongly supports this thesis (Gamer, 1972). The extremely belated conversion to democracy of the British colonial rulers of Hong Kong is surely the most bare-faced hypocrisy. Indonesia grew as a nation under nearly four centuries of Dutch dictatorship. African nations inherited authoritarian states over which democratic constitutions were hastily pasted at the time of independence. The Leninist version of Marxism has added its own brand of justification for authoritarianism.

As institutionalists (e.g. March and Olsen, 1989) have observed, when one regime replaces another the form of rule often remains the same. Marxist tyranny replaces Tsarist tyranny in Russia. Islamic fundamentalist tyranny replaces Shahist tyranny in Iran. The fault, according to Burnheim (1996), is in the idea of sovereignty itself which is infused with authoritarianism. Democracy was only imposed late upon the nation state after long and sometimes bloody struggle. This is not to say that sheer barbarism and the lust for power has not also played its usual part.

FIRST STEPS TO A WORLD CONSTITUTION FOR ENVIRONMENTAL AND ECOLOGICAL JUSTICE

Stretching Tully's analysis a little, we can say that the international system of economic and political governance has two institutional aspects which exist side by side in effortless contradiction. On the one hand the international system is a structure of ideas and institutions which supports a system of control and co-ordination of an expanding economic sphere: the system as a *uniform economic order*. On the other hand the international system is a loose structure of political relationships (based upon the principle of sovereignty) in which a variety of different interests interact and in which principles are debated with a view to regulation of the activities of corporations: the system as a *negotiated order*.

A system of relatively uniform nation states co-ordinated by market competition and a negotiated order *is* a system of governance. The question today, then, is not *whether* we should have world governance but *how* the world governance that already exists might be reconstituted as a *government* in the peoples' control, capable of carrying on and extending the dialectic of justice.

The immense, blind force of the productivist capitalist system cannot, we think, today be underestimated. The engine of capitalism pours forth commodities at an ever-expanding rate at the feet of the classes with the power to consume. There is no thought for the future, no thought for any but material values, no thought for anything but commodities, steered only by a narrow and distorted definition of justice. The Marxist hope that capitalism will collapse under the contradictory stresses of its own processes of valuation and devaluation have not so far been realised. Even if these stresses result in temporary collapse, we cannot be sure that what follows will be much better or even much different. We cannot afford to wait and see. *If* capitalism is to be restrained, it must be by a power of a magnitude and scope to match that of the capitalist engine itself. It must be of global scale. The slogan 'think globally, act locally' is no longer appropriate. Local action within an unchanged global order of production and governance rapidly reaches its limits. It is necessary today not only to think about the global consequences of local action, but to act to change the global context of local action: 'Think and act, globally and locally'.

World government does not require the existence of a corporate world administrative state like that of the nation state. Great power is necessary to restrain great power. But the power necessary to restrain and direct the global capitalist system must *itself* be restrained. The political virtue of the capitalist system is that it is not monolithic but plural. Within the system there are multiple sources of non-military power: legislative assemblies of local, regional and national scope, judicatures, pressure groups, political parties, even the competing production and service corporations themselves.

If a power is to be created that can restrain capitalism, it, too, must be restrained by powers of equal scope and authority: a constitution enshrining values of ecological and environmental justice, a network containing nodes of legislative power and authority.

A democratic world government must, we believe, protect the right of peoples to choose the form of 'local' (that is national, regional, etc.) government under which they live. This idea of autonomy is a western cultural artifact. Why should this culturally embedded idea become universal – that is global? More specifically, does the guarantee of autonomy *itself* radically change a local political framework of a more authoritarian type which may have been shaped by centuries of cultural practice? The answer is, of course, yes it does. But the idea of autonomy is already here in the world. If we follow Held's democratic thought experiment, once the choice is present, persons must be free to make it. This is the tragedy of Pandora and her box. As Thompson (1992: 190) observes, there is no practicable way back to the homogeneous community ruled by benign tradition. Both Tully and Held share a reverence for the principle of consent. If persons do not consent to the political framework within which they live, freedom is sacrificed and most probably 'belonging' as well. It is hardly likely that persons will feel a sense of belonging to a culture they perceive as oppressive. Once even the *possibility* of autonomy exists, the choice exists. No political culture can avoid dealing with the consequences of the existence of political choice.

In future the world will increasingly confront conflicts which cannot be resolved without some form of global constitutional framework. Nuclear testing is but one example. Others are global warming, the environmental effects of the international financial system, national (not just 'Third World') debt, and the problem of food and water for a growing world population. That framework is already in process of construction through the UN negotiated order. But the order is weak and fundamentally divided between economic and ecological aims. The constitutional framework which eventuates must be based on principles which apply to all nations and all peoples.

Burnheim (1996: 64) sums up the idea of transnational ecological democracy: 'A fully defensible democracy must be conceived as a set of institutions and procedures that secures for all human and non-human beings the best natural and social environment that can be achieved with the means available to us'. Once we are decided on the goal to be pursued, many other questions arise which take us far beyond the scope of this book. How might a global constitution be developed via the negotiated order of the United Nations? What form of popular representation should be adopted? How are citizens to be guaranteed participation in and access to institutions of cosmopolitan democracy? How are the interests and rights of the non-human natural world to be represented and advocated? What means are available to a global institution to ensure compliance with its decisions? What form of sanctions may be used in the implementation of global regulation, and how shall such sanctions be applied? How might cash flow be arranged to permit the necessary international transfers for compensation and welfare? What is the time horizon for the negotiation of a global constitution? How much time do we have left?

Such questions as these can only be slated for future discussion. But we can at least suggest some first steps. Rather than a comprehensive cosmopolitan system with general powers, a safer solution, at least in the shorter term, might be to develop a set of institutions under the mantle of the United Nations mandated to deliver ecological and environmental justice (see Low and Gleeson, 1997). A directly elected World Environment Council and an International Court of the Environment should be created. Both have already been foreshadowed. A Global Environmental Organization created under the UN with comparable authority to the World Trade Organization has been proposed (see Lipietz, 1992: 124; Esty, 1994; French, 1995: 184; Postiglione: 1996: 12). A proposal for an International Court of the Environment was put forward at the Rio Conference in 1992 by the the International Court of the Environment Foundation based in Rome (Postiglione, 1996). The proposal was also considered sympathetically by the European Community in 1993. Postiglione states: 'The call for an International Court of the Environment is justified not only by its human rights aspect but also by the strongly felt social and ethical need for *environmental justice*' (1994: 22, author's emphasis).

The Council would provide a forum for political debate and public scrutiny of environmental issues. The Court would adjudicate specific disputes involving matters affecting the environment which have a clear international dimension. These two institutions provide the mechanisms through which justice can be discursively defined in conditions in which coercion is absent. Together they would gradually acquire the legitimacy to implement such measures as are necessary to ensure the effectiveness of world environmental law.

Burnheim warns against the danger of a global sovereign state becoming totalitarian. A nexus of transnational institutions equipped with democratic authority would serve the aims of democracy: 'A group of recognised international authorities backing each other up could be powerful enough to exercise decisive sanctions even over most states in most circumstances' (Burnham, 1996). Though some issues such as global warming and nuclear testing might well come within its direct purview, the purpose of a directly elected world environment council would *not* be to make decisions on specific issues of environmental conflict. It would do two things: establish the *principles* for making such decisions and create the *institutional mechanisms* for making them. The latter would be a series of accountable authorities bringing together the interests concerned, much like the various multilateral committees set up under the aegis of the United Nations for the creation of regulatory regimes (see Young, 1994). The World Environment

Council would be a constitutional council – involved among other tasks in that of shaping its own constitution. Such a council would *not* replace the negotiated order but strengthen it.

The mode of election of the Council should almost certainly *not* be based upon the representation of existing national states, but upon peoples and communities. If, as Burnheim suggests, it is preferable that interests be represented, it must be remembered that ultimately only human interests count and not the interests of some artificial 'person in law'. Nevertheless some interests may be territorially based but some may not be. So an electoral system would have to be devised to provide for the representation of both territorial and non-territorial interests. A place could be made in this representational system for the transnational corporation – but only if the corporation itself embraced certain ethical principles. The detail of such principles might well form an early item on the agenda of the World Environment Council. One principle, however, can be stated now. Representation should be conditional upon the corporation embracing democracy. In other words the government of the corporation should be subject to the same democratic principles as the government of a state, as argued by Dahl (1985).

And what of non-human interests? Such interests will need to be represented by human advocates. A role for species advocates, or ecoadvocates in a more comprehensive sense might become part of the constitutional framework. The general aim would be to provide the authority for negotiations among directly affected parties on specific issues. The powers of a world council to decide matters itself, as opposed to those matters which would be delegated to committee for negotiation, would have to be carefully specified. As Burnheim argues, the aim is to avoid decision by 'power trading'. Negotiated regimes for specific purposes would be the norm. 'Imaginative and creative proposals tend to come out of such contexts, given a will to arrive at an optimal solution' (*ibid.*: 54).

Oran Young's analysis of the negotiation of international environmental regimes has important implications for cosmopolitan democracy. First, under cosmopolitan democracy the regulatory order governing aspects of the human–nature relationship will be negotiated among the relevant political units. This accords both with Tully's principle of consent and Held's principle of the use of force only as a last resort. Obviously, however, the possibility of enforcement by a cosmopolitan (i.e. transnational) authority of some sort will change the whole framework of negotiation in ways which cannot yet be predicted. However, to rule out force absolutely and on principle may be to remove the only possibility of success. The scale and difficulty of resolution of the problem of anthropogenic climate change may be the pivotal issue on which rests a decision by the nations of the world to move towards cosmopolitan democracy.

There may be many reasons on principled grounds for favouring cosmopolitan democracy, but the crisis of a threatening global disaster with uneven impact, together with a recurrent failure to establish an effective negotiated regime, may require the use of some kind of force. The use of force then raises the question of the authority for its use (especially in the light of past occasions on which the current United Nations system has been seen to fail in legitimising force). This in turn raises the question of democracy. Thus the matter of climate change may become the lever for the creation of a wider cosmopolitan framework. Many other regimes dealing with a wide range of problems relating to the environment could then be negotiated under the new cosmopolitan framework. This framework would itself have to be negotiated and installed by consent, but, as Young points out in the case of environmental regimes, the negotiations will focus on principles of justice under conditions of a rather thick veil of uncertainty. Rawls's initial position may turn out to have a quite practical application. Likewise Held's democratic thought experiment will be relevant.

Is the international system environmentally and ecologically just? It would appear not. But the system is in transition. Finding justice is, as always, a struggle. World government under a constitution which embodies both human rights and the rights of the non-human world is indispensible to provide a just framework for the flourishing of maximum local diversity of both human cultures and natural ecologies.

CONCLUSION

A framework of international/global governance is already well established. It seems hardly likely that history will go into reverse. Both the destructive potential of international war as a means of settling disputes and the growing interdependence among nations means that national governments will continue to seek co-operation, and national sovereignty will continue to be eroded. The questions to be addressed, therefore, are whether the currently emergent framework is a good one and whether it requires further modification.

If this seems a remarkably Eurocentric view of the history of the international system, we should remember that the ideas and institutional forms developed in Europe were not unique. There are similarities as well as differences between European political cultures and political cultures which had already developed or were developing simultaneously in human societies in other parts of the world. Moreover, as Tully insists, European culture, like all cultures, is not homogeneous. But it was European nations that embarked on the expansionary course of settlement, trade and military conquest which shaped the international system. It was European variants of political ideas, albeit interpreted and shaped by contact with other cultures, which provided the justification of this system.

It is neither universality nor local autonomy *per se* that are problematic.

Both are required. It is their particular combination which poses problems of transnational governance today: the *lack* of national autonomy coupled with the continued *existence* of national sovereignty which places states in competitive economic relations with one another. In such a situation democracy is threatened and authoritarianism prospers. Nevertheless, the fear which Tully's analysis evokes is that the establishment of a cosmopolitan order based on universal principles of justice will become as imperialistic and oppressive as the 'empire of uniformity' which was imposed upon the indigenous peoples of North America and elsewhere under European hegemony, and which now rules the transnational economic order. Held's central principles, as he himself acknowledges, are 'at the core of the modern liberal democratic project', the very same project as that of Locke (see Held, 1995b: 149). We see no reason why a cosmopolitan constitution should necessarily extinguish cultural diversity. Indeed, it should be specifically designed to reinforce and constitute it.

8

THE DIALECTIC OF JUSTICE
AND NATURE

Alethic truth, as optimally grounding reason, can be the rational cause of transformative negating agency in absenting constraints on self-emancipation, that is, on the liberation of our causal powers to flourish. For to exist is to be able to become, which is to possess the capacity for self-development, a capacity that can be fully realized only in a society founded on the principle of universal concretely singularized human autonomy in nature. This process is dialectic; and it is the pulse of freedom.

(Roy Bhaskar, 1993, *Dialectic, The Pulse of Freedom*, p. 385)

INTRODUCTION

The dialectic of justice involving humans, their societies and nature is the process of finding the truth of the human condition. In this final chapter we briefly review the nature of the dialectic, we then consider the principles suggested by a second level of political justice: the principles of a global constitution for the institutions we wish to see develop. We revisit the examples of environmental conflict with which this book began and sketch some scenarios of how they might be dealt with by global institutions. Finally we consider how a process of transformation at global level might unfold.

THE DIALECTIC OF JUSTICE AND NATURE

We are persons located in space and time, and we also belong to a species whose individual members are able to perceive how they fit in to a complex whole. We are part of humanity, and also part of nature. Since we can reflect on the whole – of humanity, society, and nature extended in space and time – we can also reflect on distributions within this whole. We can consider the political question: who (and what) gets what and where? We can further reflect imaginatively on an alternative to the actual distribution and then compare the imagined with the actual. In imagining a distribution and comparing imagined and actual we make judgements. Judgements flow from and into the imaginative process of comparing. Judgements act as the stimulus for comparison as well as reflections on comparisons. In comparing, we use judgements like 'better' or 'worse'. We can then reflect on what makes one distribution better than another. So we arrive at criteria. In relationships in which we feel a moral bond, these criteria are the bases of justice.

Having decided that one kind of distribution is better, that is fairer or more just, we can move up a step and consider how the *system* in which we live can be made to deliver that kind of distribution rather than another 'worse' distribution. We arrive at criteria of *political* justice. One way of ensuring that political justice is done is to insist that people have inalienable rights. Whatever we think of or feel towards a person, that person deserves to have her or his rights respected. A person has a right to the satisfaction of her or his basic needs.

This process is never conducted by humans in isolation. We draw on the fund of ideas provided by our history, culture and place in the world. We encounter the fact that other humans, differently placed, also go through a similar process. And they arrive at different conclusions: about the distributions regarded as better or worse, about the criteria of 'better' or 'worse', and the means for arriving at better and worse outcomes. We can then move up a further step and consider what sort of process for dealing with the fact of difference would itself be better or worse. So this is a *second level of political justice*.

Potentially, and logically, there is an infinite regress. We can go on thinking about the criteria for deciding how to decide how to decide . . . *ad infinitum*. In practice, though, it is probably safe to assume that people will reach agreement, or something close to it, at the second level of political justice. It is a converging process. This is the process of finding justice, or justice as an open-ended dialectical process.

In the first flush of triumph at the imminent collapse of communism, Fukuyama could only envisage a closed Hegelian stasis: the end of 'history':

> The struggle for recognition, the willingness to risk one's life for a purely abstract goal, the worldwide ideological struggle that called forth daring, courage, imagination, and idealism, will be replaced by economic calculation, the endless solving of technical problems, environmental concerns, and the satisfaction of sophisticated consumer demands. In the post-historical world there will be neither art nor philosophy, just the perpetual care-taking of the museum of human history.
>
> (Fukuyama, 1989: 13)

We hope that this book has put an end to such nonsense for, as Bhaskar observes, 'Process in open totalities entails that all politics are transitional, and that all causally efficacious transformative praxis is continually negating the status quo' (1993: 269). The dialectical struggle must continue as long as human beings are alive and free on this planet. But of course the struggle of ideas, social forces and nature will take new forms. We have argued throughout this book that non-human nature is morally considerable, though the work of generating from this perception moral precepts which can give rise to practical results through the construction of institutional means is still in its infancy.

The dialectic, then, contains an evolving debate about the bases of justice: in deserts, rights and needs. These bases are founded in turn upon conceptions of the self and its relations with human and non-human nature. A first level of political justice contains conceptual systems for resolving conflicts over distributions between humans and humans, and humans and non-human nature. A second level of political justice contains concepts for resolving disputes about first-level political systems. A 'foundational' conception of the universal is, we think, untenable. There is no conceivable system of ideas about justice which will resolve all conflicts in a perfectly fair manner, and against which no reasonable argument can be mounted. The truth of the human condition is stratified. At every level, principled judgement is required to resolve conflicts of principle of the level below. The dialectic must continue because, as Bhaskar says, it is 'the pulse of freedom', the human pulse.

Nevertheless, the idea of the universal itself drives the dialectic. The acceptance of relativism brings it to a halt and renders it meaningless. If all we can say to another person is, 'Your conceptions of justice are true for you, in your cultural context, but mine are true in my context', meaningful debate about justice must cease. In fact the whole idea of justice becomes meaningless. To debate about justice entails the idea that there is a *truth* to which we aspire and which we seek through communication with others. Bhaskar calls this 'alethic truth': 'the truth of, or real reasons for, or dialectical ground of, *things*, as distinct from *propositions*' (Bhaskar, 1993: 394; note that this is not a distinction between 'things' and 'persons', but between the real and the ideal). The significance of this insight cannot be underestimated. There is a real world of which we are part – outside and separate from our perceptions of it and propositions about it. The question of justice is the question of our human place in that real world. Dialectic is not just a cacophony of signs thrown around in a maelstrom of political skirmishes. Dialectic entails the development of enlarged thought about the real world and our place in it.

Science continues the search for alethic truth, though even science today is not immune from the disease of relativism. It is increasingly said that science merely serves political interests and is inserted into political struggles in support of those who can pay for research. It is not unknown for scientists to prostitute their vocation for money. It is also understandable where survival is at stake. Fortunately, however, there is still a science independent of politics. Hard evidence, facts, are still good currency. Science still embodies a genuine dialectic, a search for alethic truth. But science concerns itself almost exclusively with the reality of the non-human world. If science applies itself to humans (medical science, for example), humans are treated as things, cases of disease, cases of health, statistics, ciphers. All too often animals are also treated as things for instrumental purposes. True, scientists are sometimes deeply devoted to ethical standards. But ethics, as Laing (1982) observed, is not considered part of what science knows. Ethical standards are imposed from elsewhere, frequently from the world of politics.

Politics in much of the world has become reduced to *competitive* politics. *Deliberative* politics, in which the search for the truth of the human condition is sought, has almost disappeared. Politics is simply about winning power. Competitive politics is about personalities and signs: politicians conjure images like the 'bridge to the future', 'time for a change', never mind that the bridge leads nowhere, and the change is of personnel, not of policy. This has not happened because of the apostasy of political leadership. Political leaders like Bill Clinton and Tony Blair are thrown up by a politics which is responsive to the condition of the people who elect them. That condition is one of chronic, and sometimes acute, insecurity. When people are struggling to maintain the material standards to which they have been taught to aspire, to propose different policies is dangerously to rock the boat.

Insecurity is the negative side of competition. We are reaching a situation in the world in which competition, understood in the most limited and trivial sense, focused upon the desire for material objects, immediate gain and instant gratification, deathly to the higher ends for which humans strive, is generating a situation in which little else can be seriously considered. Unfortunately there is a positive feedback effect. The more the destruction of the environment continues, the more that resources are depleted, the more insecurity is generated, the more competition intensifies. No-one can predict the political evil in which this destructive course of world affairs will end.

The reinstatement of deliberative politics requires the relief of the competitive insecurity into which the peoples of the world have been thrown. Whatever is done must be global in scope and meet standards of distributive justice, environmental justice. But, though global, the question is not an inter*national* one. Third World poverty now exists within 'rich' nations. First World riches now exist in 'poor' nations. It is a question of distribution among people, not among nations. As we have argued, ecological justice has to be considered at the same time as environmental justice. But no progressive transformative political agenda can be be expected to receive mass support until the context of global competition is changed to increase the level of economic security worldwide.

THE SECOND LEVEL OF POLITICAL JUSTICE

The ideas we have canvassed in this book are of western origin. Like all ideas they have cultural roots. But it is our firm belief that people from different cultural traditions can enter into dialogue because cultures interpret in different ways a *common* reality. As Thompson observes:

> The problematic nature of any pronouncements about international justice is not a reason for not making them, but rather a reason for recognising that they are only a contribution to an ongoing debate which ought not to be dominated by western concerns and interests.
>
> (Thompson, 1992: 190)

Throughout, we have argued in favour of the right to need satisfaction, including *environmental* need. We argued for an enlarged conception of political justice inclusive of the rights of the non-human world, based upon extended horizons of the self. We abhorred the traffic in 'risk'. At core, the problem we identified was the inability of any but a narrow conception of justice to penetrate the world's governing institutions. Institutional transformation is required at global level in order to open global institutions to the dialectic of justice.

A first step towards such 'opening up', we said, might be a World Environment Council and an International Court of the Environment. These institutions, properly constituted, might themselves play a role in further transformation of the global system in the interests of environmental and ecological justice. However, these institutions, and any others which may be created in a future network of global agencies, must themselves be subject to principles of justice at a second political level. We propose four principles to govern the constitution of the World Environment Council and Environment Court: the ecological principle, the principle of autonomy, the principle of uncoerced discourse, and the principle of consent. These principles, in lexical order, follow from the discussion in the foregoing chapters (*cf.* the four objectives for international justice proposed by Thompson, 1992: 188).

The ecological principle

The first principle of ecological justice, as we argued in Chapter 6 is that, 'every natural entity is entitled to enjoy the fullness of its own form of life'. This principle extends the right to respect and dignity to all non-human nature. We did not accept that biospherical egalitarianism is logically acceptable. Conflict among species is a fact of existence, not least between human and non-human forms of life. In part this conflict is a natural result of mutual dependency. So acceptance of the principle of mutual dependency is also accepted as a second principle of ecological justice: 'all life forms are mutually dependent and dependent on non-life forms'. Resolving such conflicts is a matter for human moral judgement through the institutions created by humanity for such a purpose – including the World Environment Council and Environment Court.

In the second ecological principle we recognise that conflict of interest is a fact, not only of human life but of all existence. No foundational principle can be discovered which will itself resolve such conflict. There are principles which are true, but to some degree conflict with one another. All that can humanly be done is to create the means of judgement which will embody 'enlarged thought' (see Benhabib, 1992; also Rawls, 1993a: 58). Subject to the above principles, we made three distinctions which may be regarded as reasonable guidelines for judgement:

1 Life has moral precedence over non-life.
2 Individualised life-forms have moral precedence over life-forms which *only* exist as communities.
3 Individualised life forms with human consciouness have moral precedence over other life forms.

Ultimately it will be the task of the World Environment Council to adapt and modify such conceptions of moral precedence, and for the Environment Court to interpret them in acts of judgement.

The principle of autonomy

This principle, it will be recalled, was stated by Held as follows:

> Persons should enjoy equal rights, and accordingly, equal obligations in the specification of the political framework which generates and limits the opportunities available to them; that is they should be free and equal in the determination of the conditions of their own lives, so long as they do not deploy this framework to negate the rights of others.

> (Held, 1995b: 147)

This principle is of course subject to the ecological principle which recognises that 'the rights of others' extends beyond the frontier of the human. But the principle of autonomy also instantiates for humans the principle that every natural entity is entitled to enjoy the fullness of its own form of life. Humans are entitled to enjoy the *fullness* of their *human* form of life.

Modified by the ecological principle, autonomy cannot be understood simply in terms of freedom to produce, exchange and consume the commodities and services which existing power structures permit to be produced, exchanged and consumed. As we have seen in Chapter 6, the horizons of the *autos* (self) in 'autonomy' may be defined in a wider and more inclusive sense. The flourishing of the human being entails allowing for the *choice* of these wider horizons of the self. Of course such a choice cannot be imposed, or it would not be a choice. The 'expanded self' cannot in any way be legislated. All that can be done is gradually to discover a form of society in which the free choice of self becomes possible, such that the choice is not limited to the restricted, and narrowed self-picture which can alone justify the existing capitalist system. Even if in the fullness of time many people embrace versions of the 'expanded self', this will not abolish the potential conflict of interest between humanity and the rest of nature – as some deep ecologists have thought. This is why the ecological principle which extends rights to the non-human is necessary and why the question of justice must be continually in question. Nevertheless, any increase in the tendency to define the self against wider horizons will at least blunt the conflict between humanity and the non-human and make consensus a little easier to obtain.

The principle of autonomy also establishes that a *democratic* basis for the World Environment Council and other global institutions is necessary. The concept of representative democracy combines the principle of autonomy, embodied in the accountability of the representative, with a recognition of the stratified reality of human existence. Representatives must act as judges in cases of conflicting interest and conflicting principle. Two fundamentals are thereby acknowledged: the right of individual persons to determine the conditions of their own lives, and the right of persons acting collectively to change structures of power which tend to restrict human flourishing. The creation of global institutions is now the absolute pre-requisite for change of such power structures, for these power structures and steering mechanisms are themselves of global scope. They already constitute world governance.

The autonomy principle is not to be interpreted in terms of 'formal' or legal rights alone. As Doyal and Gough (1991) have argued, moral person-hood requires that all basic human needs be met. These needs include health, housing and education, and also time and intellectual resources to participate fully as a citizen. The provision for 'citizenship' may also encourage the enlargement of the boundaries of the self. As we have seen, Shue (1980; 1992) and Galtung (1994) also include the provision for basic rights (the right to need satisfaction) as a requirement of justice. It is not a distortion of Kant's conception of property and welfare to say that he too endorsed such a right as flowing from the categorical imperative and the conception of human persons as 'ends-in-themselves'. Peffer (1990: 418) argues that not only liberty must be protected but 'equal worth of liberty'. Rawls (1993a: 265) introduces the idea of 'background justice' to embrace much the same idea, though he stops short of extending it to the international sphere (see Rawls *ibid*.: 272 fn).

The principle of uncoerced discourse

Underlying the idea of discursive democracy is the principle that the people involved in making a decision should not be under any coercion whatsoever.

Inherent in real argumentation, aimed at 'reaching understanding' in Habermas's words, is the presumption that all force is absent, whether it comes from within the argument (for example *ad hominem* attacks, suppressions of fact, lies, distortions, and other rhetorical devices designed to win arguments) or externally on the participants (e.g. through pressure, bribery and threats). Only the force of the better argument is to prevail. This principle is in fact embodied, though imperfectly, in the secret ballot, parliamentary privilege, academic, judicial and bureaucratic 'tenure' of office, and freedom of the press. It is also enshrined in the view that bribery is 'corrupt'. Both the deliberations of the World Environment Council and the Environment Court must be subject to the principle of uncoerced discourse. So, too, must be the process of election of delegates to the Council and that of appointment of judges to the Court.

Of particular importance in observance of the principle of uncoerced discourse is freedom of information. This freedom includes not only public access to information already produced, but freedom to produce information. The freedom to produce must be understood not just as a negative freedom – the absence of coercion upon the producers of knowledge about the environment, but also as a positive freedom. The production of knowledge about the environment must not be starved of funds. The diversity and independence of sources of knowledge production must be maintained. Thus, it is ethically essential for schools and universities to be financially independent from the economic interests of production. It is critically important that the scientific integrity of knowledge about the human–environment relationship be protected. Integrity is best protected when a thriving and independent scientific community exists to scrutinise knowledge production, a community whose members are not beholden to commercial or state interests and where the only force prevailing is that of the better argument.

There is of course a rather important contradiction between any system of representative democracy and discursive democracy. In order to realise the principle of autonomy, delegates (for example, to the Council) must be under the coercion of accountability to their constituents. Representative democracy can be no more than a reasonable compromise between the capacity of delegates to engage in uncoerced discourse in the process of deliberation, and their capacity to embody the people's autonomy by being regularly accountable at election. Moreover, even with the best form of proportional representation, delegates may still make themselves subject to the discipline of factions and parties. If an executive body of some kind is to be formed from the Council delegates, and if there is to be majoritarian decision making, then this will divide the Council into supporters and opponents of the executive.

These and other considerations have encouraged Dryzek (1990; 1994) to look for discursive democracy in the 'public spheres' beyond the central institutions of governance (committees for negotiating settlements, public inquiries, non-government organisations, etc.). Certainly we consider the expansion of these public spheres to be highly desirable; they play a central role in a plural democracy. However, further progress in an ecologically benign politics cannot ignore the reform of the central institutions. A system of proportional representation in which delegates are elected for specified periods still seems the best institutional compromise. It does not seem altogether impossible for a deliberative, respectful and open politics which confronts state (that is nation state) power – as recommended by Dryzek, (1994: 186) to be generated within the World Environment Council, though there is not space to consider here the question of precisely how this might be achieved.

The principle of consent

The World Environment Council and the Environment Court requires the consent of peoples of different political cultures. Discourses of justice are embedded within cultures, as Habermas and the promoters of discourse ethics, as well as interculturalists like Tully recognise. Hence, it is necessary for speakers to speak *authentically* within their cultural context and acknowledge the validity of the cultural context of others. The principle of consent must, however, be stated separately from that of uncoerced discourse in order to give adequate recognition to the importance of cultural difference. The principle of autonomy demands that individuals be free to decide on the political system under which they live. But there are different valid ways in which individuals' needs can be represented collectively, consistent with the principle of autonomy (Tully, 1995: 123). Neither the establishment nor the praxis of a World Environment Council is intended to override the forms of collective representation which have developed in different national cultures, nor to impose a 'uniform constitution' over the globe.

People come to the World Council from different cultures. We would expect their concepts of justice to differ though we would also anticipate enough of an 'overlapping consensus' (in Rawls's words) to make real intercultural discourse possible. There is surely enough precedent for this in the experience of the United Nations.

Any system of principles of justice must be based upon a conception of the self. The above principles follow from an expanded self-conception inclusive of the environment of other people and nature. We reject the picture of the 'bounded' self because it is not an accurate reflection of reality, indeed it is a distortion. Nevertheless, it is a widely prevalent image. Some versions of deep ecology and Marxist thought seem to suppose that an expanded self-picture will only come into being once the basis of the world's systems of production and governance are changed. *How* those systems are to be changed is a separate matter from what will emerge once they are changed. We do not agree. The process of changing the system must *itself* conform to the principles we wish to see emerging. To espouse a course of political action involving violence (for example), would negate the very picture of self that we wish to nurture: the mature expanded self with the capacity for enlarged thought. The institutional solution we are proposing would do no more than open the way for such a picture to take hold. It would not and cannot guarantee its emergence.

Both communitarian and contractarian ideas would be put to the test. The World Environment Council would be an, albeit embryonic, expression of a global community. The mechanisms necessary to elect such a Council would themselves promote the idea of world community. We shall not here consider the practical difficulties of creating an electoral system and a voting system which would enable the Council to function as a democratically elected authority. However, today's communications technology already has the capacity to achieve this goal. For instance, the Internet is already widely used for disseminating information about the environment and about places and local communities. In Chapter 5 we mentioned the use of EcoNet to link grassroots activist organisations seeking environmental justice. The Web has been used effectively in Mexico, Bosnia and Serbia to bring pressure to bear on governments to heed the will of the people. The potential of the global Web to spread uncensored information is an immense power for human emancipation and the struggle against the still vast coercive power of the nation state. Of course information can also include disinformation, deceitful and dangerous information. The Web will be used for all kinds of power struggles. It will be used to recruit people to despair and injustice. Recall the incident in 1997 when the Web was used to recruit people to a self-mutilating and suicidal cult of salvation by intergalactic visitors. But the spatial extension of human interaction which the Web represents cannot now be abolished. All the more reason, then, for global institutions which uphold standards of truth.

The use of technology to expand the *public* sphere is certainly a task for the next century. The global community based on global interaction and information will bring dangers. The world will have to consider how to achieve a good balance between freedom and care. But unless the whole course of the Enlightenment is to be put into reverse, the freedom to access information of any kind will continue to outweigh protection of people from the consequences of their own choices. A central issue will be the accuracy, truth and objectivity of information flowing both to Council delegates and to their constituencies. Almost certainly some further auditing or 'watchdog' bodies would have to be created with full powers to oversee the freedom of access to and dissemination of information.

The deliberations of the Council constituted under the above principles would also *enact* the social contract. There is no need for a hypothetical foundational event. The persons elected to the Council would enact bargains among themselves. What these bargains would be cannot be predicted.

There is no need for a 'difference principle' based upon what free agents *might* hypothetically agree because, within the Council, free agents would *actually* engage in the process of reaching agreement. The Council would not, however, be all powerful. It would be constrained by the dispersal and separation of powers, both vertically and horizontally, among the nodes of a network of global government. Vertically, national, regional and local governments would retain many of their powers, and committees for the negotiation of environmental regimes authorised by the Council would also form a separate source of power. Horizontally, the powers of the Council would be restrained by the Environment Court, by the Council's own constitutional principles, and by a variety of auditing bodies.

Postiglione (1996: 48) identifies the following premises for the constitution of the Environment Court: (a) a new legal basis must be created for the court – he suggests a new 'framework convention' of the world's states under the UN; (b) the court must function as a supranational authority with decision-making powers; (c) the court must be accessible to individuals and non-government organisations as well as states; (d) the court would have a body of independent and non-removable judges, would conduct public hearings, would found decisions on specified grounds, would have the power to implement controls related to prevention of environmental damage and would have the power to adopt emergency measures and order economic sanctions and compensation.

THE PRAXIS OF THE SECOND LEVEL

Let us now revisit the examples discussed in Chapter 1, the disposal of the Brent Spar oil rig, nuclear testing by France in the Pacific and the mining operations at Ok Tedi. How might the course of these events have been different had the global institutions we recommend been in place? Of course, once we consider these concrete issues a large number of further questions become apparent which, at this point, we can only skate over.

A major concern of the World Environment Council (WEC) would, at first, be to bring into being a comprehensive environmental regulatory regime extending worldwide which would cover all aspects of production. The aim would be to anticipate the environmental costs of production and ensure that these costs are fully felt at the *source* of production. Environmental costs must be shifted from the community, including its non-human members, back to the generators of those costs. Thus, the costs of production of oil at sea (in the Brent Spar case) would be made to include the costs of disposal of all the waste products, including obsolete production plant, with minimal environmental impact. The precise standards defining 'minimal impact' would be the subject of negotiation.

The typical mechanism for achieving such regulation and arriving at fair standards of 'minimal impact' would be the 'negotiated regime' of a kind already occurring under the United Nations system. The WEC would have ratified the membership of a committee to negotiate such a regime. Let us call it the Intercommunal Committee for the Regulation of Oil Production (ICROP). Producer corporations might be directly represented in such a forum provided that the managements of the corporations represented were subject to internal democracy inclusive of all the corporation's workers. Once agreed by ICROP, the rules of production would require the consent of the WEC. Such consent having been given, the regime would come into force and a range of penalties would apply to breaches, with the ultimate sanction of exclusion from the world trading community. Questions of breach of standards, and application of sanction would be the province of the International Court of the Environment.

Because oil production is prone to environmentally damaging accidents, a committee would certainly have been created early in the life of the first Council to negotiate such a regulatory regime. It is quite possible, however, that this regulatory regime would take a long time to negotiate. The WEC would have to decide on which environmental dangers require urgent action. Anthropogenic climate change and protection of the ozone layer are examples of such matters. But many other matters require a long process of inquiry, negotiation and consensus building and this is likely to be a slow process when measured against a human life span. At global level, however, a politically sustainable regime based on the best available knowledge is preferable to one rapidly arrived at – and as rapidly overturned.

In the course of negotiation, incidents such as the attempted disposal of the Brent Spar might well occur. It would be for ICROP to decide on an interim regime pending final agreement which would provide some basic regulatory principles. If a producer corporation still decided to challenge these principles, objectors could apply to the International Court of the Environment for an injunction to stop the dumping while argument was heard and evidence adduced on both sides. The assumption we are making here is that a forensic process with adequate rules of evidence is more likely to lead to a discursively just outcome than a political battle in which information is thrown into a political arena to manipulate public opinion. The court would make a decision based on the evidence and could either permit the action (dumping) or order the corporation to desist from or change its course of action.

Following the nuclear testing in the Pacific in 1973, the French Government repudiated its recognition of the International Court of Justice at The Hague. Here is an example of a State flouting international opinion and conducting a process which self-evidently does extreme damage to the environment and generates a large area of risk. To prevent this happening, a regulatory regime would have to be created either to govern the testing of nuclear devices in the environment or, more widely to govern the global armaments industry. It seems likely that, as with the oil industry, the WEC would give this matter high priority. However, interfering with the right of a state to its own military defence is a matter of great sensitivity concerning state sovereignty. Such interference may, in fact, only become possible following a more radical reform of the United Nations system along lines suggested by Held (1995b) and Archibugi (1995).

Certainly, however, the very existence of a World Environment Council wielding the authority of the peoples of the world is a powerful instrument for the containment of national hubris. Its existence would act as a deterrent. However, should a nation state decide to go ahead with a course of action with extreme dangers to the environment, then that nation could be brought before the Environment Court by another national government or by a non-government organisation. The court could order an immediate injunction to stop the damaging operation pending examination by the court. The court could also order the nation to desist and, in the event of the nation continuing its damaging operation, the court could order appropriate sanctions. Ultimately, the sanction of total exclusion from world trade would apply.

The dispute concerning the Ok Tedi mining operation in Papua New Guinea is only one of a number of disputes which would be the concern of the World Environment Council and Environment Court. Many of these disputes involve both states and corporations. In these disputes the coercive power of the state is typically brought in to support the economic power of the corporation in the context of gross economic dependency upon a structure of global debt. As the examples of the Ogoni people in Nigeria and that of the Panguna mine on the island of Bougainville demonstrate, these disputes can easily degenerate into violence and armed struggle. Already military force itself is becoming 'privatised', no longer the exclusive property of the nation state. Private armies are already being hired out to corporations to enforce their will and protect their property. It is only a matter of time before these armies become absorbed into the command structure of the corporations themselves. The Ok Tedi case is instructive in that it *was* brought before a court and has so far avoided a violent outcome. But the matter of environmental justice, let alone ecological justice, was scarcely part of the final settlement.

Let us suppose that the WEC institutes a forum for the creation of a regime of environmental regulation covering the world's mountains: the Intercommunal Committee for Mountain Conservation (ICMoC). As Denniston has argued, mountains must be moved – up the global agenda:

Mountain people urgently need equitable and sustainable human development. Today, the impacts of the ecological, economic and social challenges they face extend far beyond the 2 billion people who live downstream. Their environments are integrally connected to the lowlands through the movements of water, animals, people and products. To restore and maintain the quality of life of mountain residents and the health of mountain ecosystems, global efforts towards sustainable development must be integrated with efforts to sustain mountain peoples and places.

(Denniston, 1995: 57)

We have argued that the primary concern should be with environmental and ecological justice and not merely with sustainable development. Through ICMoC a regime for the protection and reasonable development of mountain areas would be negotiated consistent with the demands of environmental and ecological justice. Clearly, however, such negotiation would raise a question which extends far beyond the scope of the committee to resolve. The question is this: is it fair to prevent a poor nation from exploiting the resources within its territory in order to bring the material standards of life of the population of the nation as a whole up to those of the developed world? The question might initially be posed as one of compensation. How might the population of the nation in question be compensated for witholding maximum development? Here also the rights of the mountain peoples themselves must be entered. The rights of the population must never be considered in aggregate, for the 'population' is differentiated by class and property and the mountain peoples may be among the least powerful. ICMoC might arrive at a working compromise but the wider question will be posed of the justice of a highly differentiated and stratified world order, both within and between nations. The structures of national debt are just one manifestation of this stratified order. Others have proposed solutions (e.g James Tobin – see French, 1995: 185; Lipietz, 1992: 112 *et seq.*; Jacobs, 1991: 184). It would be for the WEC to pursue this issue through specialist economic and constitutional fora which it would itself institute. In this way WEC would itself become a forum for discussion of the further reform of the international system in the direction of global justice.

In the meantime, the International Court of the Environment would be the appropriate court for handling disputes over major development projects such as that of Ok Tedi. In the Environment Court questions of environmental and ecological justice would be addressed and judgements made, taking into account these wider considerations. With widespread popular respect for the authority of the Court's decisions, resort to violence on the part of local actors would be unlikely. How could such popular support be secured? Only, perhaps, with recognition that the Court is born of the need to shield the world from a great evil which is not just a historical event but an ever-present condition caused by the past absence of restraint.

The praxis described above would not be a final solution to the problematic of the environment. We emphasise that we regard such praxis as a first step towards the transformation of a production system which is out of control and heading for planetary destruction. The critique we have advanced in this book envisages not merely the re-regulation of the economy but its democratisation. Our aim is to see production and consumption profoundly reshaped so that needs are met fairly and ecological health and global integrity is maintained. We look towards a decommodified society in which the objects and means of production are determined socially and ecologically, based on principles of justice – a break from both capitalism and state socialism. If there is a more deliberative link between the parliamentary role of the WEC and its regulatory powers, extended eventually beyond externality control to proscription of production of such forms as toxic waste and maleficent weaponry, then the movement towards a gradual democatisation of production would have begun. We have proposed a democratic mechanism to start this process, but the process itself must proceed dialectically and observing the principles of justice we advocate above.

THE FUTURE OF THE DIALECTIC

The experience of the twentieth century suggests strongly that transformation of a politico-economic system only occurs when the tensions within it build up to the point of crisis. Change at that point can be sudden and comprehensive. Experience also tells us that ensuing developments will be shaped by ideas which have been nurtured over many years, ideas which before the time of crisis have been dismissed as irrelevant, inappropriate or dangerous. Instances include majoritarian democracy, the dictatorship of the proletariat, the welfare state, the Fascist corporate state and the wave of privatisation and deregulation which swept through the developed world from the 1980s. As Cockett (1996) shows, ideas which became hegemonic in the 1980s and 1990s were developed and nurtured far from the mainstream in such organisations as the Mont Pélérin Society founded by Hayek – itself modelled on the Fabian Society.

We cannot know what ideas will take hold following the next crisis. We do know, with reasonable certainty, that there will be such a crisis. It is extremely important, therefore, to debate the ethical implications of ideas which now seem outlandish and unnecessary. We need to prepare for the transformation of the politico-economic system beyond the crisis to ensure that the transformation is benign. We have proposed that the democratisation of world governance would be a step towards a benign and politically sustainable transformation whose eventual form cannot be clearly seen from where we stand.

The democratisation of world government in pursuit of ecological and environmental justice will not occur until the tensions within the existing system of production and governance build up to the point of crisis. The potential sources of crisis within the capitalist mode of production are numerous and well known. Capitalism faces continual and growing environmental crises as local communities resist the sacrifice of their environments for profits, and an ecological crisis as aggressive competition consumes the health and integrity of the planet.

The emergent politics may take surprising directions. To take one possible scenario, if the current consensus among climatologists is correct, an increase in global warming caused by human industry will begin to produce detectable and attributable localised effects. Not least among these effects will be a rise in sea level. The IPCC report (1996: 388) states: 'The "best estimate" (for the scenario) IS92a is that sea level will rise by 49 cm by the year 2100, with a range of uncertainty of 20–86 cm' (see Figure 8.1). The residential settlement patterns of the most favoured, the gold coasts of the world, sprawl around the beaches and bays and secluded islands. How is the value of such an expensive sprawl to be protected? Will those who have been forced to live in property away from the desired coastal ecotone agree to pay the taxes necessary to protect the property of the fortunate? Will the value of the coast itself be unaffected by ugly walls and dykes?

The actors next century who will support action to prevent global warming may be not only 'Greens' and ecophilosophers but also those actors and institutions within the existing structures of economic and political power who stand to lose most: the coastal property industry and residents, the tourism industry and the insurance industry. Even so it is not perhaps the question of who is going to pay for the sea walls and dykes, or how they are going to pay, so much as the continuing symbol they represent of humanity's negligence. That negligence will be dramatised by the increased and sometimes cataclysmic flooding of the deltas and low-lying lands occupied by the very poor, the gradual obliteration of vast tracts of land from which the occupants can hardly afford to flee let alone build walls. Consider Bangladesh.

The future fissures and alignments of global politics will change. New coalitions will form. If there is to be change in the world's systems of governance, the change will not emerge from the sort of politics we see today – with minority nationally based green movements appealing for public support for a programme of revolutionary change, and green pressure groups (even global groups) engaging in skirmishes against the 'capitalist enemy'. Certainly that element of environmentalism will continue to exist, but the Green movement will enormously expand its horizons. The spreading of the Green movement into hegemonic form necessarily involves fragmentation and pluralisation. 'The Greens' should maintain their position, but even their most 'extreme' wings should not vilify those who want to make alliances with power-holders and elites. A *purist* green politics which is also broad enough to constitute a new hegemony is impossible. In the new hegemony 'Green' will fade to be absorbed into the 'common sense' of all parties, movements and interest groups. This is not to advocate a 'pale green' capitalism but many different groups with different material interests and different value systems share a real interest in environmental and ecolo-

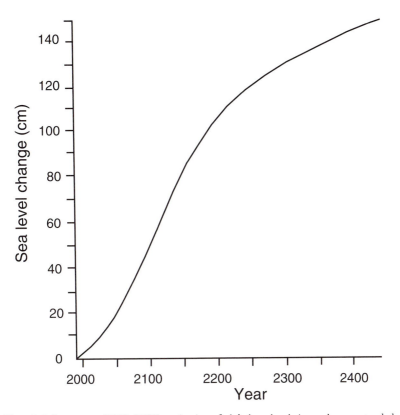

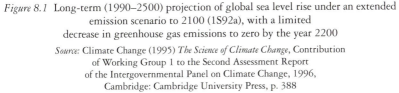

Figure 8.1 Long-term (1990–2500) projection of global sea level rise under an extended emission scenario to 2100 (1S92a), with a limited decrease in greenhouse gas emissions to zero by the year 2200

Source: Climate Change (1995) *The Science of Climate Change*, Contribution of Working Group 1 to the Second Assessment Report of the Intergovernmental Panel on Climate Change, 1996, Cambridge: Cambridge University Press, p. 388

gical justice and the democratisation of production. A successful broad 'green' (small 'g') movement will actively seek out and identify shared inter-ests with a view to negotiating less destructive ways of living on the Earth.

If a new hegemony is to be formed it cannot be done by force. Thus Ernst von Weizsäcker rejects 'ecotyranny'. How can ecotyranny be prevented?

Firstly, ecological transformation should start in good time, while there is still plenty of room for manoeuvre and before the urgency of the situation subordinates all other considerations. Secondly, we should promote those environmental policy instruments which besides being effective leave room for individual response, and we should refuse those which are repressive. Finally, we should start to concern ourselves now

211

with liberties which will have to be protected once the exigencies of the Century of the Environment become everyday realities.

(Weizsäcker, 1994: 210)

We should not assume for a moment that global warming and destruction of the ozone layer are the only planetary changes we will need to worry about. The next century will undoubtedly produce knowledge of new and dangerous anthropogenic transformations of the Earth. The effects of genetic engineering, radiation from communication systems, agricultural monocultures, particulate and gas pollution from automobile exhausts, desertification, the loss of tropical rainforest, the use, storage and transport of hazardous chemicals, the hazards of armaments production, the disposal of the vast amounts of nuclear waste, and many other unknown future risks we are even now incurring will bring new dangers to the public view. The value of the irreplaceable local environments and ecosystems continually being destroyed will increasingly be weighed against the value of the products thereby produced: how much of the natural environment is a mobile phone worth?

But this knowledge must be accompanied by knowledge of the political institutions with which to negotiate the conflict which will undoubtedly ensue from change. The more agency is mobilised in favour of environmental values, the more there will be agency mobilised to resist such values. We have already seen resistance begin in the United States in 'the war against the Greens' (Helvarg, 1994). The early part of the next century may be marked not by the dawn of environmental consensus but by a massive reaction against environmentalism. Under the present world system of economy and governance, many people will suffer material hardship from ecological regulation. Jobs are still tied to 'economic growth'. Material well-being is still tied to jobs. Unless we can find ways of breaking that nexus, ecological destruction will continue to be supported as the lesser of two evils by many working people. Such a break cannot be achieved by any single nation state acting alone.

CONCLUSION

At the end of the century a time of great weariness has settled on the polity worldwide. The hope which attended the birth of socialism and democracy in the early twentieth century has melted into air. Post-modernism, ostensibly opposing systems of domination, has aided the withering of interest in history and in social projects. Everything – places, people, information, environments, time, space, even history itself, is homogenised into a regime of marketable commodities for highly unequal consumption. As the economic crisis of capitalism unfolds, there seems no way out of the polarised world in which immense poverty and social and environmental breakdown coexists with immense riches and power. It is a 'plutopia' of mutual insecurity in which individuals anxiously face the future without hope of mutually shaping it. The security of the few is temporarily built upon the daily insecurity of the many. But in the long term even that security fades as the production system changes the conditions of life for all. The ecological crisis forces all people to face their real conditions of life and their true relations with their kind – both human and non-human. Beneath the surfaces, celebrated in so-called post-modernity, behind the happy-face mask of fast food and commodity consumption, lies actually existing capitalism dedicated to the acceptance of this polarised world as the best of all possible worlds.

This predatory world is natural enough, but it is inhuman. It is not the morality of the animal world we want to see upheld but the human. We oppose the politics of indifference. We want to see hope rekindled, community reinstated, human history regained, the future reconsidered. Environmental and ecological justice are ultimately about security for people across the globe, for places, for environments and for the planet. The challenge of the new century, the challenge of ecological and environmental justice is nothing less than the transformation of the global institutions of governance, the reinstatement of democracy at a new level, the democratisation of both production and its regulation.

BIBLIOGRAPHY

Abbey, E. (1986) 'Letter to the Editor', *Bloomsbury Review*, April–May, p. 4.
—— (1988) 'Immigration and liberal taboos', in *One Life at a Time Please*, New York: Henry Holt & Co.
Ackroyd, P., Anderson, T. and Hartley, P. (1991) *Environmental Resources and the Market Place*, Sydney: Allen & Unwin.
Adeola, F. O. (1994) 'Environmental hazards, health and racial inequity in hazardous waste distribution', *Environment and Behavior*, 26/1: 99–126.
Allende, I. (1988) *Of Love and Shadows*, London: Black Swan Books.
Alston, D. (ed.) (1990) *We Speak for Ourselves: Social Justice, Race and the Environment*, Washington, DC: Panos Institute.
Altvater, E. (1993) *The Future of the Market, An Essay on the Regulation of Money and Nature after the Collapse of 'Actually Existing Socialism'*, London: Verso.
—— (1994) 'Ecological and economic modalities of time and space', in O'Connor, M. (ed.), *Is Capitalism Sustainable? Political Economy and Political Ecology*, New York: Guilford Press, pp. 76–90.
Anderson, P.W., Arrow, K.J. and Pines, D. (eds) (1988) *The Economy as a Complex Evolving System*, Santa Fe Institute Studies in the Science of Complexity, vol. 5, Redwood City, CA: Addison-Wesley.
Anderson, T. and Leal, D. (1991) *Free Market Environmentalism*, San Francisco: Pacific Research Institute for Public Policy.
Apel, K.-O. (1990) 'The problem of a universalistic macroethics of co-responsibility', in Griffioen, S. (ed.) *What Right Does Ethics Have?*, Amsterdam: VU University Press, pp. 23–40.
Archibugi, D. (1995) 'From the United Nations to cosmopolitan democracy', in Archibugi, D. and Held, D. (eds) *Cosmopolitan Democracy, An Agenda for a New World Order*, Cambridge: Polity Press, pp. 121–62.
Aristotle (1976 edn) *Nicomachean Ethics*, Harmondsworth: Penguin.
Armour, A. (1991) 'The siting of locally unwanted land uses: towards a cooperative approach', *Progress in Planning*, 35 (1): 1–74.
Attfield, R. (1983) 'Christian attitudes to nature', *Journal of the History of Ideas*, 44 (3): 369–86.
Badcock, B. (1984) *Unfairly Structured Cities*, Oxford: Blackwell.
Bahro, R. (1982) *Socialism and Survival*, London: Heretic Books.
—— (1986) *Building the Green Movement*, London: GMP.
Baier, A. (1992) 'The need for more than justice', *Canadian Journal of Philosophy*, suppl., vol. 13.
Bailey, J. (1988) *Pessimism*, London: Routledge.
Baker, J. (1987) *Arguing for Equality*, London: Verso.

Baker, R. (1996) 'The vast frontier', *The Bulletin*, Sydney, 21 May, pp. 46–7.

Banerjee, B.N. (1986) *Bhopal Gas Tragedy – Accident or Experiment?*, New Delhi: Paribus.

Barry, B. (1989) *A Treatise on Social Justice, Volume 1: Theories of Justice*, Berkeley: University of California Press.

Barry, J. (1996) 'Sustainability, political judgement and citizenship: connecting green politics and democracy', in Doherty, B. and de Geus, M. (eds) *Democracy and Green Political Thought*, London: Routledge, pp. 115–31.

Bateman, I.J. and Turner, R.K. (1994) 'Valuation of the environment, methods and techniques: the contingent valuation method', in Turner, R.K. (ed.) *Sustainable Environmental Economic and Management: Principles and Practice*, London: Belhaven, pp. 120–91.

Bateson, G. (1979) *Mind and Nature, A Necessary Unity*, New York: E.P. Dutton.

Bauman, Z. (1993) *Postmodern Ethics*, Oxford: Blackwell.

—— (1995) *Life in Fragments, Essays in Postmodern Morality*, Oxford: Blackwell.

Beck, U. (1992) *Risk Society, Towards A New Modernity* (trans. Mark Ritter), London: Sage.

—— (1995) *Ecological Politics in An Age of Risk*, Cambridge: Polity Press.

Beckerman, W. (1974) *In Defence of Economic Growth*, London: Jonathan Cape.

—— ([1975] 1990) *Pricing for Pollution: Market Pricing, Government Regulation, Environmental Policy*, 2nd edn, London: Institute of Economic Affairs.

—— (1994) ' "Sustainable Development": Is it a Useful Concept?', *Environmental Values*, 3: 191–209.

Been, V. (1993) 'What's fairness got to do with it? Environmental justice and the siting of locally undesirable land uses', *Cornell Law Review*, 78: 1,001–85.

—— (1994) 'Locally undesirable land uses in minority neighborhoods: disproportionate siting or market dynamics?' *Yale Law Review*, 103: 1,383–422.

Bell, D. (1976) *The Cultural Contradictions of Capitalism*, New York: Basic Books.

Bellah, R., Madsen, R., Sullivan, W.M., Swidler, A. and Tipton, S.M. (1985) *Habits of the Heart, Individualism and Commitment in American Life*, New York: Harper & Row.

Benhabib, S. (1992) *Situating the Self, Gender, Community and Postmodernism in Contemporary Ethics*, Cambridge: Polity Press.

Bennett, J. and Block, W. (eds) (1991) *Reconciling Economics and the Environment*, Perth: Australian Institute for Public Policy.

Bentham, J. (1843) 'Principles of the Civil Code', in *Collected Works*, published posthumously under the superintendence of his executor, Edinburgh: William Tate.

Benton, T. (1993) *Natural Relations: Ecology, Animal Rights and Social Justice*, London: Verso.

Berger, P.L. (1987) *The Capitalist Revolution, Fifty Propositions about Prosperity, Equality and Liberty*, Aldershot: Wildwood House.

Bhaskar, R. (1993) *Dialectic, The Pulse of Freedom*, London: Verso.

Birnie, P. (1992) 'International environmental law: its adequacy for present and future needs', in Hurrell, A. and Kingsbury, B. (eds) *The International Politics of the Environment*, Oxford: Clarendon Press, pp. 51–84.

Blomley, N.K. (1985) 'The Shops Act (1950): the politics and the policing, *Area*, 17: 25–33.

—— (1989) 'Law and the local state: enforcement in action', *Transactions of the Institute of British Geographers*, NS, 13: 199–210.

Blowers, A. (1996) 'Environmental policy – ecological modernisation or the risk society?', unpublished paper, copy obtained from author, The Open University, Walton Hall, Milton Keynes, UK.

Boerner, C. and Lambert, T. (1995) 'Environmental injustice', *Public Interest*, Winter, pp. 61–82.

Bogard, W. (1989), *The Bhopal Tragedy: Language, Logic, and Politics in the Production of a Hazard*, Boulder, CO: Westview Press.

Bookchin, M. (1980) *Toward an Ecological Society*, Montreal: Black Rose Books.

—— (1982) *The Ecology of Freedom, The Emergence and Dissolution of Hierarchy*, Palo Alto, CA: Cheshire Books.

—— (1990) *Remaking Society, Pathways to a Green Future*, Boston, MA: South End Press.

—— (1995a) *From Urbanization to Cities, Towards a New Politics of Citizenship*, London: Cassell.

—— (1995b) *Re-enchanting Humanity, A Defense of the Human Spirit against Anti-humanism, Misanthropy, Mysticism and Primitivism*, London: Cassell.

Bottomore, T., Harris, L., Kiernan, V.G. and Miliband, R. (eds) (1991) *A Dictionary of Marxist Thought*, 2nd edn, Oxford: Blackwell.

Boyne, G. A. and Powell, M. (1991) 'Territorial justice in Britain: a review of theory and evidence', *Political Geography Quarterly*, 10: 263–81.

Brechin, S.R. and Kempton, W. (1994) 'Global environmentalism: a challenge to the postmaterialism thesis?' *Social Science Quarterly*, 75/2: 245–69.

Brennan, A. (1988) *Thinking About Nature: An Investigation of Nature, Value and Ecology*, London: Routledge.

Brian, A. W. (1990) 'Positive feedbacks in the American economy', *Scientific American*, February, pp. 92–9.

Bromley, D.W. (ed.) (1995) *The Handbook of Environmental Economics*, Oxford: Blackwell.

Brubaker, E. (1995) *Property Rights in the Defense of Nature*, London: Earthscan.

Brundtland Report (1987) *Our Common Future*, Oxford: Oxford University Press (World Commission on Environment and Development).

Bryant, B. and Mohai, P. (eds) (1992) *Race and the Incidence of Environmental Hazards*, Boulder, CO: Westview Press.

Bullard, R. (1990a) *Dumping in Dixie*, Boulder, CO: Westview Press.

—— (1990b) 'Ecological inequalities and the new South: black communities under siege', *The Journal of Ethnic Studies*, 17 (4): 101–15.

—— (1992a) 'Unequal environmental protection: incorporating environmental justice in decision-making', paper presented at the Resources for the Future Conference on Setting National Environmental Priorities: the EPA risk-based paradigm and its alternatives, Annapolis, MD (R. Bullard: Sociology Department, University of California, Riverside, CA 92521).

—— (1992b) 'Environmental blackmail in minority communities', in Byrant, B. and Mohai, P. (eds) *Race and the Incidence of Environmental Hazards: A Time for Discourse*, Boulder, CO: Westview Press, pp. 82–95.

—— (1993a) 'Anatomy of environmental racism and the Environmental Justice Movement', in Bullard, R. (ed.) *Confronting Environmental Racism: Voices from the Grassroots*, Boston, MA: Southend Press.

—— (1993b) 'Waste and racism, a stacked deck?' *Forum for Applied Research and Public Policy*, Spring edn, pp. 29–35.

—— (1993c) 'Anatomy of environmental racism', in Hofrichter, R. (ed) *Toxic Struggles: The Theory and Practice of Environmental Justice*, Philadelphia, PA: New Society, pp. 25–35.

Bullard, R. and Wright, B.H. (1990) 'Toxic waste and the African American Community', *The Urban League Review*, 13 (1–2): 67–75.

Burnheim, J. (1996) 'Power-trading and the environment', in Mathews, F. (ed.) *Ecology and Democracy*, London: Frank Cass, pp. 49–65.

Burton, J. (1979) *Deviance, Terrorism and War, The Process of Solving Unsolved Social Problems*, New York: St Martins Press.

Callicott, J.B. (1985) 'Intrinsic value, quantum theory and environmental ethics', *Environmental Ethics*, 7: 257–75.

Capra, F. (1982) *The Turning Point: Science, Society and The Rising Culture*, New York: Simon & Schuster.

Carson, R. (1962) *Silent Spring*, Cambridge, MA: Riverside Press.

Carter, F.W. and Turnock, D. (1993) *Environmental Problems in Eastern Europe*, London: Routledge.

Carver, S. and Openshaw, S. (1992), 'A geographic information systems approach to locating nuclear waste disposal sites', in Clark, M., Smith, D. and Blowers, A. (eds) *Waste Location: Spatial Aspects of Waste Management, Hazards, and Disposal*, London: Routledge, pp. 105–27.

Cassese, A. (1986) *International Law in a Divided World*, Oxford: Clarendon Press.

Castells, M. (1979) *City, Class and Power*, London: Macmillan.

Charnovitz, S. (1994) 'NAFTA's environmental significance', *Environment*, March, pp. 42–5.

Charvet, J. (1995) *The Idea of An Ethical Community*, Ithaca, NY: Cornell University Press.

Childers, E. and Urquhart, B. (1994) 'Renewing the United Nations system', Development Dialogue 1, Dag Hammarskjöld Foundation/Ford Foundation.

Chisholm, A.G. and Moran A.J. (1993) *The Price of Preservation*, Melbourne: Tasman Institute.

Christoff, P. (1996) 'Ecological modernisation, ecological modernities', *Environmental Politics*, 5 (3): 476–500.

Clark, M. and Smith, D. (1992) 'Paradise lost? Issues in the disposal of waste', in Clark, M., Smith, D. and Blowers, A. (eds) *Waste Location: Spatial Aspects of Waste Management, Hazards, and Disposal*, London: Routledge, pp. 1–11.

Cockett, R. (1996) *Thinking the Unthinkable: Think Tanks and the Economic Counter-Revolution, 1931–1983*, London: HarperCollins.

Cohen, G.A. (1986) 'Self-ownership, world-ownership and equality', in Lucash, F.S. (ed.) *Justice and Equality Here and Now*, Ithaca, NY and London: Cornell University Press, pp. 108–35.

Commoner, B. (1972) *The Closing Circle*, New York: Bantam Books.

Cookes, T. (1995) 'Mururoa may be leaking radiation', *The Age*, Melbourne, 12 September, p. 37.

Cox, K.R. (1973) *Conflict, Power and Politics in The City*, New York: McGraw-Hill.

—— (1979) *Location and Public Problems*, Oxford: Blackwell.

Cropper, M.L. and Oates, W.E. (1992) 'Environmental economics: a survey', *Journal of Economic Literature*, XXX, June, pp. 675–740.

Cummiskey, D. (1996) *Kantian Consequentialism*, New York and Oxford: Oxford University Press.

Curtis, S. (1989) *The Geography of Public Welfare Provision*, London: Routledge.

Cutter, S. (1995) 'Race, class and environmental justice', *Progress in Human Geography*, 19 (1): 111–22.

Dahl, R.A. (1985) *A Preface to Economic Democracy*, Berkeley and Los Angeles: University of California Press.

Daly, M. (1996) 'Australia's toxic trade', *The Age*, Melbourne, 13 September, p. A9.

Davies, B.P. (1968) *Social Needs and Resources in Local Services*, London: Michael Joseph.

Davy, B. (1996) 'Fairness as compassion: towards a less unfair facility siting policy', *Risk: Health, Safety and Environment*, 7: 99–108.

Dear, M. (1977) 'Spatial externalities and locational conflict', in Massey, D.B. and Batey, P.W.J. (eds) *London Papers in Regional Science 7*, London: Pion.

—— (1992) 'Understanding and overcoming the NIMBY syndrome', *Journal of the American Planning Association*, 58: 288–99.

Dear, M. and Taylor, S.M. (1982) *Not on Our Street*, London: Pion.

Dear, M., Wolch, J. and Wilton, R. (1994) 'The human service hub concept in human services planning', *Progress in Planning*, 42 (3): 174–271.

Denniston, D. (1995) 'Sustaining mountain peoples and environments', in Brown, L. (ed.) *State of the World 1995*, London: Earthscan, pp. 38–57.

Devall, B. (1986) 'A spanner in the woods', interview with David Foreman, *Simply Living*, 2/12: 3–4 (Australia).

—— (1990) *Simple in Means, Rich in Ends, Practicing Deep Ecology*, Salt Lake City: Peregrine Smith Books.

Diamond, S. (1981) *In Search of the Primitive*, New Brunswick, NJ: Transaction Books.

Dicken, P. and Lloyd, PE. (1981) *Modern Western Society*, London: Harper & Row.

Dobson, A. (1990) *Green Political Thought, An Introduction*, London: Unwin Hyman.

Doyal, L. and Gough, I. (1991) *A Theory of Human Need*, London: Macmillan.

Drengson, A. (1980) 'Shifting paradigms: from the technocratic to the person–planetary', *Environmental Ethics*, 3: 221–40.

—— (1988) Review of Devall and Sessions, *Deep Ecology: Living as if Nature Mattered*, in *Environmental Ethics*, 10: 83–9.

Dryzek, J. (1987) *Rational Ecology, Environment and Political Economy*, Oxford: Blackwell.

—— (1990) *Discursive Democracy: Politics, Policy and Political Science*, Cambridge: Cambridge University Press.

—— (1994) 'Ecology and discursive democracy: beyond liberal capitalism and the administrative state', in O'Connor, M. (ed.) *Is Capitalism Sustainable? Political Economy and the Politics of Ecology*, New York and London: Guilford Press.

—— (1996) 'Political and ecological communication', in Mathews, F. (ed.) *Ecology and Democracy*, London and Portland, OR: Frank Cass.

Duncan, J.W. and Shelton, W.C. (1978) *Revolution in United States Government Statistics*, Washington, DC: US Department of Commerce, Office of Federal Statistical Policy and Standards.

Dworkin, R. (1983) 'Comment on Narveson: in defense of equality', *Social Philosophy and Policy*, 1: 24–35.

Eckersley, R. (1992) *Environmentalism and Political Theory, Towards An Ecocentric Approach*, London: UCL Press (University College London).

—— (1995) 'Markets, the state and the environment: an overview', in Eckersley, R. (ed.) *Markets, the State and the Environment*, Melbourne: Macmillan, pp. 7–45.

—— (1996) 'Greening liberal democracy, the rights discourse revisited', in Doherty, B. and de Geus, M. (eds) *Democracy and Green Political Thought*, London: Routledge, pp. 212–36.

The Economist (1993) *Pocket World in Figures*, London: The Economist in association with Hamish Hamilton; published London: Penguin Books.

Edwards, R. (1995a), 'Dirty tricks in a dirty business', *New Scientist*, 18 February, pp. 12–13.

—— (1995b), 'Leaks expose plan to sabotage waste treaty', *New Scientist*, 18 February, p. 4.

Elkington, J. (1987) *The Green Capitalists*, London: Victor Gollancz.

Elster, J. (1982) 'Sour grapes – utilitarianism and the genesis of wants', in Sen, A.K. and Williams, B.A.O. (eds) *Utilitarianism and Beyond*, Cambridge: Cambridge University Press, pp. 219–38.

—— (1983) 'Exploitation, freedom and justice', *Nomos*, 26: 277–304.

—— (1985) 'Rationality, morality and collective action', *Ethics*, 96: 136–55.

Engels, F. ([1876]1995) 'The part played by labour in the transition from ape to man', in Redclift, M. and Woodgate, G., *The Sociology of the Environment,* vol. I, Aldershot: Elgar.

Esty, D.C. (1994) 'The case for a global environmental organization', in Kenen, P.B. (ed.) *Managing the World Economy: Fifty Years after Bretton Woods*, Washington, DC: Institute for International Economics.

Etzioni, A. (1993) *The Spirit of Community, Rights, Responsibilities and The Communitarian Agenda,* New York: Crown Inc.

Fagan, R.H. and Webber, M. (1994) *Global Restructuring: The Australian Experience*, Melbourne: Oxford University Press.

Feinberg, J. (1970) *Doing and Deserving, Essays in the Theory of Responsibility*, Princeton, NJ: Princeton University Press.

Feldman, S. (1992) 'Introduction', in Spinoza, B., *Ethics* (trans. S. Shirley), Indianapolis, IN: Hackett, pp. 1–20.

Ferry, L. ([1992] 1995) *The New Ecological Order* (trans. Carol Volk), Chicago: University of Chicago Press (originally published as *Le Nouvel Ordre Ecologique, L'arbre, l'animal et l'homme*: Paris: Bernard Grasset).

Fisher, R. and Ury, W. (1981) *Getting to Yes*, Boston, MA: Houghton Mifflin.

Fox, W. (1984) 'Deep ecology, a new philosophy for our time?', *The Ecologist*, 14 (5/6): 194–200.

—— (1990) *Toward A Transpersonal Ecology*, Boston, MA: Shambhala.

Frankel, B. (1987) *The Post-Industrial Utopians*, Cambridge: Polity Press.

Fraser, N. (1987) 'What's critical about critical theory? The case of Habermas and Gender', in Benhabib, S. and Cornell, D. (eds) *Feminism as Critique, Essays on the Politics of Gender in Late-Capitalist Societies*, Cambridge: Polity Press, pp. 31–56.

French, H.F. (1995) 'Forging a new global partnership', in Brown, L. (ed.) *State of the World 1995*, London: Earthscan Publications for Worldwatch Institute.

Fukuyama, F. (1989) 'The end of history', *The National Interest*, 16: 3–18.

Galtung, J. (1994) *Human Rights in Another Key*, Cambridge: Polity Press.

Gamer, R.E. (1972) *The Politics of Urban Development in Singapore*, Ithaca, NY and London: Cornell University Press.

Gare, A. (1995) *Postmodernism and the Environmental Crisis*, London: Routledge.

Giddens, A. (1990) *The Consequences of Modernity*, Stanford, CA: Stanford University Press.

Gillespie, E. and Schellhas, B. (1994) *Contract with America*, New York: Times.

Gilligan, C. (1982) *In a Different Voice, Psychological Theory and Women's Development*, Cambridge, MA: Harvard University Press.

Gilligan, C., Ward, J.V. and Taylor, J.M., with Bardige, B. (1988) *Mapping the Moral Domain, A Contribution of Women's Thinking to Psychological Theory and Education*, Cambridge, MA: Center for the Study of Gender, Education and Human Development, Harvard University.

Gleeson, B.J. (1995) 'The commodification of planning consent in New Zealand', *New Zealand Geographer*, 51 (1): 42–8.

—— (1996) 'Justifying justice', *Area*, 28 (2): 229–34.

Gleeson, B.J. and Memon, P.A. (1994) 'The NIMBY Syndrome and Community Care Facilities: A Research Agenda for Planning', *Planning Practice and Research,* 9 (2), 105–18

Goldblatt, D., Held, D., McGrew, A. and Perraton, J. (forthcoming) *What is Globalization? Concepts, Theories and Evidence,* Cambridge, UK: Polity Press.

Goldman, B. (1996) 'What is the future of environmental justice?', *Antipode,* 28 (2): 122–41.

Goodin, R.E. (1988) *Reasons for Welfare,* Princeton, NJ: Princeton University Press.

—— (1995) *Utilitarianism as a Public Philosophy,* Cambridge: Cambridge University Press.

Gore, A. (1993) *Earth in the Balance: Ecology and the Human Spirit,* New York: Plume.

Gorz, A. (1994) *Capitalism, Socialism, Ecology* (trans. C. Turner), London: Verso.

Greenpeace International (1994) Asia toxic trade patrol, unpublished bulletin, Amsterdam: Greenpeace International.

—— (1996) *Asia Toxics Bulletin,* Amsterdam: Greenpeace International, June.

Greenpeace New Zealand (1994), *Zero by 2000: Time for Zero Discharge of Toxic Pollution in New Zealand,* Auckland: Greenpeace New Zealand.

Gregg, S.R., Mulvey, J.M. and Wolpert, J. (1988) 'A stochastic planning system for siting and closing public service facilities', *Environment and Planning A,* 20: 83–98.

Grundmann, R. (1991) *Marxism and Ecology,* Oxford: Clarendon Press.

—— (1993) 'Introduction' in Guignon, C.B. (ed.) *The Cambridge Companion to Heidegger,* Cambridge: Cambridge University Press, pp. 1–41.

Gusman, S. (1981) 'Policy dialogue', *Environmental Comment,* November, pp. 14–16.

Haas, P.M., Keohane, R.O. and Levy, M.A. (1993) Institutions for the Earth, Sources of Effective Environmental Protection, Cambridge, MA: MIT Press.

Habermas, J. (1984) *The Theory of Communicative Action,* London: Heinemann.

—— (1990) *Moral Consciousness and Communicative Action,* Cambridge: Polity Press.

Hajer, M.A. (1995) *The Politics of Environmental Discourse: Ecological Modernisation and the Policy Process,* Oxford: Oxford University Press.

Hampshire, S. (1951) *Spinoza,* London: Faber & Faber.

Hardin, G. (1968) 'The tragedy of the commons', *Science,* 162: 1243–48.

Hardin, R. (1988) *Morality within the Limits of Reason,* Chicago, IL: University of Chicago Press.

Harter, P. J. (1982) 'Negotiating regulations: a cure for malaise', *Georgetown Law Journal,* 71: 1–118.

Hartley, T.W. (1995) 'Environmental justice: an environmental civil rights value acceptable to all world views', *Environmental Ethics,* Fall, 17: 277–8.

Harvey, D. (1972) *Society, the City and the Space-Economy of Urbanism,* Association of American Geographers, Commission on College Geography, resource paper 18.

—— (1973) *Social Justice and the City,* London: Edward Arnold.

—— (1981) 'The urban process under capitalism: a framework for analysis', in Dear, M. and Scott, A.J. (eds) *Urbanization and Urban Planning in Capitalist Societies,* London: Metheun, pp. 91–122.

—— (1982) *The Limits to Capital,* Oxford: Basil Blackwell.

—— (1992) 'Social justice, postmodernism and the city', *International Journal of Urban and Regional Research,* 16: 588–601.

—— (1993a) 'The nature of the environment: the dialectics of social and environmental change', *Socialist Register,* pp. 1–51.

—— (1993b) 'Class relations, social justice and the politics of difference', in Keith, M. and Pile, S. (eds) *Place and the Politics of Identity,* London: Routledge, pp. 41–66.

—— (1996) *Justice, Nature and the Geography of Difference*, Oxford: Basil Blackwell.

Hay, A.M. (1996) 'Concepts of equity, fairness and justice in geographical studies', *Transactions of the Institute of British Geographers*, 20 (4): 500–8.

Hayek, F. A (1944) *The Road to Serfdom*, Sydney: Dimocks Book Arcade.

—— (1960) *The Constitution of Liberty*, London: Routledge & Kegan Paul.

—— (1976) *Law, Legislation and Liberty*, London: Routledge & Kegan Paul.

—— (1979) *Social Justice, Socialism and Democracy, Three Australian Lectures*, Sydney: The Centre for Independent Studies.

Häyry, M. (1994) *Liberalism and Applied Ethics*, London: Routledge.

Hayward, T. (1994) *Ecological Thought: An Introduction*, Cambridge: Polity Press.

Hazarika, S. (1987) *Bhopal: The Lessons of a Tragedy*, Harmondsworth: Penguin.

Heiman, M.K. (1996) 'Race, waste, and class: new perspectives on environmental justice', *Antipode*, 28 (2): 111–21.

Held D. (1995a) 'Democracy and the international order', in Archibugi, D. and Held, D. (1995) *Cosmopolitan Democracy, an Agenda for A New World Order*, Cambridge: Polity Press, pp. 96–120.

—— (1995b) *Democracy and the Global Order*, Stanford, CA: Stanford University Press.

Held, D. and McGrew, A.G. (1993) 'Globalization and the liberal democratic state', *Government and Opposition*, 28/2.

Heller, A. (1974) *The Theory of Need in Marx*, London: Allison & Busby.

—— (1987) *Beyond Justice*, Oxford and New York: Blackwell.

Helvarg, D. (1994) *The War Against the Greens*, San Francisco: Sierra Club Books.

—— (1995) 'Legal assault on the environment', *The Nation*, 30 January, pp. 126–30.

Hirst, P. and Thompson, G. (1996) *Globalization in Question: The International Economy and the Possibilities of Governance*, Cambridge: Polity Press.

Hoban, T.M. and Brooks, R.O. (1987) *Green Justice: The Environment and the Courts*, Boulder, CO: Westview Press.

Hobbes, T. ([1651] 1929) *Leviathan*, Oxford: Clarendon Press.

Hofrichter, R. (1993) 'Introduction', in Hofrichter, R. and Gibbs, L. (eds), *Toxic Struggles: The Theory and Practice of Environmental Justice*, Philadelphia, PA: New Society Publishers, pp.1–10.

Horwitz, T.M. (1993), 'International environmental protection after the GATT tuna decision: a proposal for a United States reply', *Case Western Reserve Journal of International Law*, 25 (1): 55–77.

Hurrell, A. and Kingsbury, B. (eds) (1992) *The International Politics of the Environment*, Oxford: Clarendon Press.

Intergovernmental Panel on Climate Change (IPCC) (1996) *The Science of Climate Change*, Cambridge, UK: Cambridge University Press (edited by J.T. Houghton, L.G. Meira Filho, B.A. Callander, N. Harris, A. Kattenberg and K. Maskell).

Jackson, R. (1995) 'A mine of disinformation', *The Bulletin*, Sydney, 23 January, p. 13.

Jacobs, M. (1991) *The Green Economy, Environment, Sustainable Development and the Politics of the Future*, Boulder, CO and London: Pluto Press.

—— (1995) 'Sustainability and 'the market': a typology of environmental economics', in Eckersley, R. (ed.) *Markets, the State and the Environment*, Melbourne: Macmillan, pp. 46–72.

Janelle, D.G. and Millward, H.A. (1976) 'Locational conflict patterns and urban ecological structure', *Tijdschrift voor Econ. en Soc. Geografie*, 67: 102–13.

Janicke, M. (1990) *State Failure: The Impotence of Politics in Industrial Society*, Cambridge: Polity Press.

Jary, D. and Jary, J. (1991) *Dictionary of Sociology*, London: HarperCollins.

Johnsen, H. (1992), 'The adequacy of the current response to the problem of contaminated sites', *Environmental and Planning Law Journal*, 9 (4): 230–46.

Johnson, S.P. and Corcelle, G. (1989) *The Environmental Policy of the European Communities*, London, Dordrecht and Boston: Graham & Trotman/Martinus Nijhof.

Johnston, R.J. (1984) *City and Society*, 2nd edn, London: Hutchinson.

Johnston, R.J., Gregory, D. and Smith, D. (1994) *The Dictionary of Human Geography*, Oxford: Blackwell.

Jones, G., Robertson, A., Forbes, J. and Hollier, G. (1990), *Dictionary of Environmental Science*, London: HarperCollins.

Kant, I. ([1790] 1892) *Kant's Kritik of Judgement*, London: Macmillan.

—— ([*c.* 1780] 1963) *Lectures on Ethics* (trans. L. Infield), New York: Harper & Row.

—— ([*c.* 1798] 1991) *The Metaphysics of Morals* (trans. M. Gregor), Cambridge: Cambridge University Press.

Kapuscinski, R. (1994) *Imperium*, London: Granta Books.

Kearns, D. (1992) 'A theory of justice – and love; Rawls on the family', in Kymlicka, W. (ed.) *Justice in Political Philosophy Vol 2*, Aldershot: Elgar, pp. 480–85.

Kelsey, J. (1995), *The New Zealand Experiment: A World Model for Structural Adjustment?*, Auckland: Auckland University Press.

Kemp, R. (1990), 'Why Not In My BackYard? A radical interpretation of public opposition to the deep disposal of radioactive waste in the United Kingdom', *Environment and Planning A*, 22: 1,239–58.

Keohane, R.O., Haas, P.M. and Levy, M.A. (1993) 'The effectiveness of international environmental institutions', in Haas, P.M., Keohane, R.O. and Levy, M.A. (eds) *Institutions for the Earth, Sources of Effective Environmental Protection*, Cambridge, MA: MIT Press, pp. 3–24.

Kersting, W. (1992) 'Politics, freedom and order: Kant's political philosophy', in Guyer, P. (ed.) *The Cambridge Companion to Kant*, Cambridge: Cambridge University Press, pp. 342–66.

Keynes, J.M. ([1926] 1931) 'The end of laissez-faire', in *Essays in Persuasion*, London: Macmillan.

King, Y. (1990) 'Healing the wounds: feminism, ecology and the nature/culture dualism', in Diamond, I. and Orenstein, G.F. (eds) *Reweaving the World, The Emergence of Ecofeminism*, San Francisco: Sierra Club Books, pp. 106–21.

Knox, P.L. (1975) *Social Well-being: A Spatial Perspective*, Oxford: Clarendon Press.

—— (1982) 'Residential structure, facility location and patterns of accessibility', in Cox, K.R. and Johnston, R.J. (eds) *Conflict, Politics and the Urban Scene*, New York: St Martin's Press, pp. 62–87.

—— (1995) *Urban Social Geography: An Introduction*, 3rd edn, Harlow, UK: Longman.

Komarov, B. (1980) *The Destruction of Nature in the Soviet Union*, White Plains, New York: M.E. Sharpe Inc.

Krall, F.R. (1994) *Ecotone: Wayfaring on the Margins*, New York: State University of New York Press.

Kymlicka, W. and Norman, W. (1994) 'Return of the citizen: a survey of recent work on citizenship theory', *Ethics*, 104: 352–81.

Laing, R.D. (1982) *The Voice of Experience*, Harmondsworth: Pelican Books.

Lake, R.W. (1996) 'Volunteers, NIMBYs, and environmental justice: dilemmas of democratic practice', *Antipode*, 28 (2): 161–74.

Lake R.W. and Disch L. (1992) 'Structural constraints and pluralist contradictions in hazardous waste regulation', *Environment and Planning A*, 24: 663–81.

Lasch, C. (1978) *The Culture of Narcissism*, New York: Norton.

Lash, S. (1993) 'Reflexive modernization: the aesthetic dimension', *Theory, Culture and Society*, 10 (1): 1–23.

LeGuin, U. (1986) *Always Coming Home*, Toronto and New York: Bantam Books.

Leopold, A. (1949) *A Sand County Almanac*, Oxford: Oxford University Press.

Levy, M.A., Keohane, R.O. and Haas, P.M. (1993) 'Improving the effectiveness of international environmental institutions', in Keohane, R.O. and Levy, M.A. (eds) *Institutions for the Earth, Sources of Effective Environmental Protection*, Cambridge, MA: MIT Press, pp. 397–426.

Light, A. and Katz, E. (eds) (1996) *Environmental Pragmatism*, London: Routledge.

Lipietz, A. (1992) *Towards A New Economic Order*, Cambridge: Polity Press.

—— (1996) 'Geography, ecology, democracy', *Antipode*, 28/3: 219–28.

Livingston, J. (1984) 'The dilemma of the deep ecologist', in Everndon, N. (ed.) *The Paradox of Environmentalism*, Toronto, Canada: Faculty of Environmental Studies, York University, pp. 61–72

Lober, D.J. (1995) 'Resolving the siting impasse: modeling social and environmental locational criteria with a geographic information system', *Journal of the American Planning Association*, 61 (4): 482–95.

Lovejoy, D. (1996) 'Limits to growth?', *Science and Society*, 60 (3): 266–78.

Lovins, A., Lovins, H. and von Weizsäcker, E.-U. (1997) *Factor Four*, London: Earthscan Publications (forthcoming).

Low, N.P. (1991) *Planning, Politics and the State*, London: Unwin Hyman.

Low, N.P. and Gleeson, B.J. (1997) 'Justice in and to the environment: ethical uncertainties and political practices', *Environment and Planning A*, 29: 21–42.

Lukes, S. (1985) *Marxism and Morality*, Oxford: Oxford University Press.

Lyons, N.P. (1988) 'Two perspectives on self, relationships and morality', in Gilligan, C., Ward, J.V. and Taylor, J.M. with Bardidge, B. (eds) *Mapping the Moral Domain, A Contribution of Women's Thinking to Psychological Theory and Education*, Cambridge, Mass.: Centre for the Study of Gender, Education and Human Development, Harvard University and Harvard University Press, pp. 21–48.

Lyotard, J.-F. ([1979] 1984) *The Postmodern Condition: A Report on Knowledge* (trans. G. Bennington and B. Massumi), Minneapolis: University of Minnesota Press.

McCarthy, T. (1984) 'Introduction', in J. Habermas, *The Theory of Communicative Action*, London: Heinemann.

McDonell, G. (1991) 'Toxic waste management in Australia: why did policy reform fail?' *Environment*, 33 (6): 10–13, 33–9.

MacIntyre, A. (1981) *After Virtue, A Study in Moral Theory*, Notre Dame, IN: University of Notre Dame Press.

Macy, J. (1987) 'Faith and ecology', *Resurgence*, July/August, pp. 18–21.

Magraw, D. (1994) 'NAFTA's repercussions: is green trade possible?' *Environment*, March, 14–45.

Mandelker, D.R. (1981) *The Environment and Equity: A Regulatory Challenge*, New York: McGraw-Hill.

Manhattan Borough Board (1993) *Review of the Two-Year Application of the Criteria for the Siting of City Facilities*, New York: Manhattan Borough President.

Mann, M. (1986) *The Sources of Social Power*, vol. 1, Cambridge: Cambridge University Press.

March, J.G. and Olsen, J.P. (1989) *Rediscovering Institutions, The Organizational Basis of Politics*, New York: Free Press.

Marshall, T.H. (1950) *Citizenship and Social Class, and Other Essays*, Cambridge: Cambridge University Press.

Marx, K. ([written 1844 and first published 1959] 1977) *Economic and Philosophic Manuscripts of 1844*, Moscow: Progress.

—— ([written 1857/8 and first published 1953] 1973) *Grundrisse (Foundations of the Critique of Political Economy, Rough Draft)*, Harmondsworth: Pelican Books.

Marx, K. and Engels, F. ([1845] 1957) *The Holy Family, or A Critique of Critical Critiques*, London: Lawrence & Wishart.

Massam, B. (1980) *Spatial Search: Applications to Planning Problems in the Public Sector*, Oxford: Pergamon.

—— (1993) *The Right Place: Shared Responsibility and the Location of Public Facilities*, New York: Longman.

Massey, D. (1984), *Spatial Divisions of Labour*, London: Macmillan.

Mathews, F. (1991) *The Ecological Self*, London: Routledge.

Mead, G.H. ([1932]1980) *The Philosophy of The Present* (ed. A.E. Murphy), Chicago, IL: The University of Chicago Press.

Mead, L. (1986) *Beyond Entitlement, The Social Obligations of Citizenship*, New York: Free Press.

Medvedev, Z. (1992) 'The global impact of the Chernobyl accident five years after', in Stewart, J. M. (ed.) *The Soviet Environment: Problems, Policies and Politics*, Cambridge: Cambridge University Press, pp. 174–96.

Merchant, C. (1992) *Radical Ecology: The Search for a Livable World*, New York: Routledge.

—— (1996) *Earthcare: Women and The Environment*, London: Routledge.

Midgley, M. (1992) 'Towards a more humane view of the beasts?', in Cooper, D.E. and Palmer, J.A. (eds) *The Environment in Question, Ethical and Global Issues*, London: Routledge, pp. 28–36.

Mies, M. and Shiva, V. (1993) *Ecofeminism*, Melbourne: Spinifex.

Mill, J.S. (1971) *Utilitarianism, Liberty, and Representative Government*, London: J.M. Dent.

Miller, D. (1976) *Social Justice*, Oxford: Clarendon Press.

Mingione, E. and Morlicchio, E. (1993) 'New forms of urban poverty in Italy: risk path models in the North and South', *International Journal of Urban and Regional Research*, 17/3: 413–27.

Mises, L. von ([1949] 1963) *Human Action: A Treatise on Economics*, Chicago, IL: Contemporary Books.

Mohai, P. and Bryant, B. (1992) 'Environmental injustice: weighing race and class as factors in the distribution of environmental hazards', *University of Colorado Law Review*, 63: 921–32.

Mol, A. (1995) *The Refinement of Production: Ecological Modernization Theory and the Chemical Industry*, Utrecht: Van Arkel.

—— (1996) 'Ecological modernisation and institutional reflexivity: environmental reform in the late Modern Age', *Environmental Politics*, 5 (2): 302–23.

Moore, G.E. (1903) *Principia Ethica*, Cambridge: Cambridge University Press.

Nader, R. (1993) 'Introduction: free trade and the decline of democracy' in Nader, R., Mander, J., Wallach, L., Brown, E.G. and Lee, T. (eds) *The Case against 'Free Trade', GATT, NAFTA and The Globalization of Corporate Power*, Berkely and San Francisco, CA: North Atlantic Books and Earth Island Press, pp. 1–11.

Naess, A. (1984) 'Intuition, intrinsic value and deep ecology, Arne Naess replies', *The Ecologist*, 14 (5/6): 201–3.

—— (1989) *Ecology, Community and Lifestyle* (trans. and ed. David Rothenberg), Cambridge: Cambridge University Press.

Neale, W.C. (1982) 'Language and economics', *Journal of Economic Issues*, 16/2: 355–69.

New York Department of City Planning (1991) *The Fair Share Criteria: A Guide for City Agencies*, New York: City of New York.

Nielsen, K. (1979) 'Radical egalitarian justice: justice as equality', *Social Theory and Practice*, 5 (2).

—— (1980) 'Capitalism, socialism and justice', *Social Praxis*, 7: 3–4.

—— (1988) 'Arguing about justice: Marxist immoralism and Marxist moralism', *Philosophy and Public Affairs*, 17/3: 212–34.

Nisbet R. (1974), *The Social Philosophers, Community and Conflict in Western Thought*, London: Heinemann.

Nozick, R. (1974) *Anarchy, State and Utopia*, Oxford: Blackwell.

—— (1981) *Philosophical Explanations*, Cambridge, MA: Bellknap Press.

—— (1989) *The Examined Life*, New York: Simon & Schuster.

O'Connor, J. (1994) 'Is sustainable capitalism possible?', in O'Connor, M. (ed.) *Is Capitalism Sustainable? Political Economy and Political Ecology*, New York: Guilford Press, pp. 152–75.

O'Connor, M. (ed.) (1994) *Is Capitalism Sustainable? Political Economy and Political Ecology*, New York: Guilford Press.

O'Hare, M., Bacow, L. and Sanderson, D. (1983) *Facility Siting and Public Opposition*, New York: Van Norstrand Reinhold.

Okin, S.M. (1989) *Justice, Gender and the Family*, New York: Basic Books.

—— (1992) 'Reason and feeling in thinking about justice', in Kymlicka, W. (ed.) *Justice in Political Philosophy*, vol. 2, Aldershot: Edward Elgar, pp. 540–60.

Ophuls, W. (1973) 'Leviathan or oblivion', in Daly, H. E. (ed.) *Toward a Steady State Economy*, San Francisco: Freeman, pp. 215–30.

O'Sullivan, A. (1993) 'Voluntary auctions for noxious facilities: incentives to participate and the efficiency of siting decisions', *Journal of Environmental Economics and Management*, 25: 12–26.

Parsons, H.L. (1977) *Marx and Engels on Ecology*, London: Greenwood Press.

Pearce, F. and Tombs, S. (1993), 'US capital versus the Third World: Union Carbide and Bhopal', in Pearce, F. and Woodiwiss, M. (eds) *Global Crime Connections: Dynamics and Control*, Toronto: University of Toronto Press, 187–211.

Peet, R. (1991) *Global Capitalism: Theories of Societal Development*, London: Routledge.

Peffer, R.G. (1990) *Marxism, Morality and Social Justice*, Princeton, NJ: Princeton University Press.

Pepper, D. (1993) *Eco-socialism, From Deep Ecology to Social Justice*, London: Routledge.

—— (1996) *Modern Environmentalism, An Introduction,* London: Routledge.

Petmesidou, M. and Tsoulovis, L. (1994) 'Aspects of the changing political economy of Europe: welfare state, class segmentation and planning in the postmodern era', *Sociology,* 28/2: 419–519.

Pinch, S. (1979) 'Territorial justice in the city: a case study of the social services for the elderly in Greater London', in Herbert, D. and Smith, D. (eds) *Social Problems and the City*, Oxford: Oxford University Press.

—— (1985) *Cities and Services*, London: Routledge & Kegan Paul.

Pirie, G.H. (1983) 'On spatial justice', *Environment and Planning A*, 15: 465–73.

Pitkin, H. (1967) *The Concept of Representation*, Berkeley and Los Angeles: University of California Press.

—— (1972) *Wittgenstein and Justice*, Berkeley: University of California Press.

Plato (1955 edn) *The Republic* (trans. H.D.P. Lee), Harmondsworth: Penguin Books.

Plotkin, S. (1987) *Keep Out: The Struggle for Land Use Control*, Berkeley: University of California Press.

Plumwood, V. (1993) *Feminism and the Mastery of Nature*, London: Routledge.

Poggi, G. (1978) *The Development of the Modern State*, London: Hutchinson.

Pollis, A. (1992) 'Human rights in liberal, socialist and Third World perspective', in Claude, R.P. and Weston, B.H., *Human Rights in the World Community*, Philadelphia: University of Pennsylvania Press, pp. 146–56.

Popper, F.J. (1981) 'Siting LULUs', *Planning* (American Planning Association), April, 47/4: 12–16.

—— (1992) 'The Great LULU Trading Game', *Planning*, May, pp. 15–17.

Postiglione, A. (1994) *The Global Village Without Regulations: Ethical, Economical, Social and Legal Motivations for an International Court of the Environment*, Florence: Giunti.

—— (1996) *The Global Environmental Crisis: The Need for an International Court of the Environment*, report of the International Court of the Environment Foundation, Florence: Giunti.

Pulido, L. (1994) 'Restructuring and the contraction and expansion of environmental rights in the United States', *Environment and Planning A*, 26: 915–36.

—— (1996) 'A critical review of the methodology of environmental racism research', *Antipode*, 28 (2): 142–59.

Rawls, J. (1971) *A Theory of Justice*, Cambridge, MA: Harvard University Press.

—— (1982) 'Social unity and primary goods', in Sen, A.K. and Williams, B.A.O. (eds) *Utilitarianism and Beyond*, Cambridge: Cambridge University Press, pp. 159–86.

—— (1993a) *Political Liberalism*, New York: Columbia University Press.

—— (1993b) 'The law of peoples', *Critical Inquiry*, 20: 36–70.

Regan, T. (1983) *The Case for Animal Rights*, Berkeley: University of California Press.

Regenstein, L. (1985) 'Animal rights, endangered species and human survival', in Singer, P. (ed.) *In Defence of Animals*, Oxford: Basil Blackwell, pp. 118–34.

Reiss, H. (ed.) (1970) *Kant, Political Writings* (trans. H.B. Nisbet), Cambridge: Cambridge University Press.

Renner, M. (1995) 'Budgeting for disarmament', in Brown, L. (ed.) *State of the World, 1995*, London: Earthscan Publications, pp. 150–69.

Reynolds, D.R. and Honey, R. (1978) 'Conflict in the location of salutary facilities', in Cox, K.R. (ed) *Urbanization and Conflict in Market Societies*, London: Methuen.

Reznichenko, G. (1989) *The Aral Catastrophe*, cited in Kapuscinski, R. (1995), *Imperium*, London: Granta Books.

Rocheleau, D., Thomas-Slayter, B. and Wangari, E. (eds) (1996) *Feminist Political Ecology, Global Issues and Local Experiences*, London: Routledge.

Rodman, J. (1977) 'The liberation of nature', *Inquiry*, 20: 83–121.

Rose, J.B. (1993) 'A critical assessment of New York City's Fair Share Criteria', *Journal of the American Planning Association*, 59 (1): 97–100.

Roszak, T. (1978) *Person/Planet*, Garden City, New York State: Doubleday.

Ruff, T. (1995) 'The ongoing risks in French tests', *The Age*, Melbourne, 4 September, p. 12.

Ryle, M. (1988) *Ecology and Socialism*, London: Rodins.

Sale, K. (1980) *Human Scale*, New York: Coward, Cann & Geoghegen.

—— (1985) *Dwellers in the Land: The Bio-regional Vision,* San Francisco: Sierra Club.

226

Salmon, G. (1991) 'Resource management legislation in New Zealand: the first step to a green economy?', in Ackroyd, P., Anderson, T. and Hartley, P. (eds) *Environmental Resources and the Market Place*, Sydney: Allen & Unwin, pp. 57–84.

Sandel, M. (1982) *Liberalism and the Limits of Justice*, Cambridge: Cambridge University Press.

—— (1984) 'Justice and the good', in M. Sandel (ed.) *Liberalism and its Critics*, Oxford: Basil Blackwell, pp. 159–76.

Sandler, B. (1994) 'Grow or die: Marxist theories of capitalism and the environment', *Rethinking Marxism*, 7 (2): 38–57.

Sarokin, D.J. and Schulkin, J. (1994) 'Environmental justice: co-evolution of environmental concerns and social justice', *The Environmentalist*, 14/2: 121–29.

Schumpeter, J.A. (1943) *Capitalism, Socialism and Democracy*, London: Allen & Unwin.

Schwartzman, D.W. (1996) 'Introduction', *Science and Society*, 60 (3): 261–5.

Seager, J. (1993) 'Creating a culture of destruction: gender, militarism and the environment', in Hofrichter, R. (ed.) *Toxic Struggles: The Theory and Practice of Environmental Justice*, Philadelphia, PA: New Society, pp. 58–66.

Seldon, A. (1990) 'Preface to the first edition' [1975], in Beckerman, W., *Pricing for Pollution: Market Pricing, Government Regulation, Environmental Policy*, 2nd edn, London: Institute of Economic Affairs, pp. 6–10.

Self, P. (1970) 'Nonsense on stilts: the futility of Roskill', *Political Quarterly*, July, 41: 249–60.

Sen, A.K. and Williams, B.A.O. (eds) (1982) 'Introduction', in *Utilitarianism and Beyond*, Cambridge: Cambridge University Press, pp. 1–22

Sessions, G. (1989) 'Ecocentrism and global ecosystem protection', *Earth First!*, 21: 26–8.

Sheng, C.L. (1991) *A New Approach to Utilitarianism, A Unified Theory and its Application to Distributive Justice*, Dordrecht: Kluwer Academic Publishers.

Sher, G. (1987) *Desert*, Princeton, NJ: Princeton University Press.

Shiva, V. (1991) *The Violence of The Green Revolution: Third World Agriculture, Ecology and Politics*, London: Zed Books.

Shrivastava, P. (1992), *Bhopal: Anatomy of a Crisis*, 2nd edn, London: Paul Chapman.

Shue, H. (1980) *Basic Rights, Subsistence, Affluence and U.S. Foreign Policy*, Princeton, NJ: Princeton University Press.

—— (1992) 'The unavoidability of justice', in Hurrell, A. and Kingsbury, B. (eds) *The International Politics of the Environment*, Oxford: Clarendon Press, pp. 373–97.

Simon, J. and Kahn, H. (1984) *The Resourceful Earth: A Response to Global 2000*, Oxford: Blackwell.

Singer, P. (1975) *Animal Liberation*, New York: Avon Books.

—— (1979) *Practical Ethics*, Cambridge: Cambridge University Press.

—— (1985) (ed.) *In Defence of Animals*, Oxford: Basil Blackwell.

Sjöstedt, G. (ed.) (1993) *International Environmental Negotiation*, Newbury Park, CA and London: Sage.

Skelton, R. (1995) 'Paradise privatised', *The Age*, Melbourne, 23 September, p. A15.

Smith, D. and Blowers, A. (1992) 'Here today, there tomorrow: the politics of hazardous waste transfer and disposal', in Clark, M., Smith, D. and Blowers, A. (eds) *Waste Location: Spatial Aspects of Waste Management, Hazards, and Disposal*, London: Routledge, pp. 208–26.

Smith, D.M. (1975) *Patterns in Human Geography*, Harmondsworth: Penguin.

—— (1977) *Human Geography: A Welfare Approach*, London: Edward Arnold.

—— (1979) *Geography, Inequality and Society*, Cambridge: Cambridge University Press.

Smith, N. (1984) *Uneven Development*, Oxford: Blackwell.

—— (1994) 'Uneven development', in Johnston, R.J., Gregory, D. and Smith, D. (eds) *The Dictionary of Human Geography*, Oxford: Blackwell, pp. 648–51.

Sontheimer, S. (ed.) (*c.* 1991) *Women and the Environment: A Reader, Crisis and Development in the Third World*, London: Earthscan Publications.

Soper, K. (1991) 'Postmodernism, subjectivity and the question of value', *New Left Review*, 186: 120–8.

Spinoza, B. ([1677] 1992) *Ethics* (trans. S. Shirley), Indianapolis, IN: Hackett.

Spretnak, C. and Capra, F. (1986) *Green Politics: The Global Promise*, Santa Fe, NM: Bear.

Stretton, H. (1976), *Capitalism, Socialism and the Environment*, Cambridge: Cambridge University Press.

Sundararajan, P.T.S. (1996) 'From Marxian ecology to ecological Marxism, *Science and Society*, 60 (3): 360–79.

Swyngedouw, E. (1992) 'The Mammon quest. "Glocalisation", interspatial competition and the monetary order: the construction of new scales', in Dunford, M. and Kafkalas, G. (eds) *Cities and Regions in the New Europe*, London: Bellhaven Press, pp. 39–67.

Szabo, M. (1993) 'New Zealand's poisoned paradise', *New Scientist*, 31 July, pp. 29–33.

Taylor, C. (1979) *Hegel and Modern Society*, New York: Cambridge University Press.

—— (1984) 'Hegel: history and politics', in Sandel, M. (ed.) *Liberalism and Its Critics*, Oxford: Basil Blackwell, pp. 177–99.

—— (1986) 'The nature and scope of distributive justice', in Lubarsh, F. (ed.) *Justice and Equality Here and Now*, Ithaca, NY and London: Cornell University Press, pp. 34–67.

—— (1991) *The Ethics of Authenticity*, Cambridge, MA: Harvard University Press.

Thompson, J. (1992) *Justice and World Order, A Philosophical Inquiry*, London: Routledge.

Tilly, C. (ed.) (1975) *The Formation of National States in Western Europe*, Princeton, NJ: Princeton University Press.

Tokar, B. (1987) *The Green Alternative*, San Pedro, CA: R. and E. Miles.

Tully, J. (1995) *Strange Multiplicity, Constitutionalism in an Age of Diversity*, Cambridge: Cambridge University Press.

Turner, B.S. (1993) 'Outline of a theory of human rights', *Sociology*, 27/3: 489–512.

Turner, R.K. (1994) 'Postscript: future prospects', in Turner, R.K. (ed.) *Sustainable Environmental Economic and Management: Principles and Practice*, London: Belhaven, pp. 383–6.

United Church of Christ (1987) *Toxic Wastes and Race in the United States*, a National Report on the Racial and Socioeconomic Characteristics of Communities with Hazardous Waste Sites, New York: United Church of Christ Commission for Racial Justice.

—— (1991) 'Principles of environmental justice', *Proceedings of the First National People of Color Environmental Leadership Summit*, Washington, DC (United Church of Christ, Commission for Racial Justice, 105 Madison Avenue, New York, NY 10016; also 475 Riverside Drive Suite 1950, New York, NY 10115).

United Nations (1993) *The World Environment, 1972–1992 – Two Decades of Challenge*, New York: United Nations Publications.

United States Environmental Protection Agency (USEPA) (1992) *Environmental Equity: Reducing Risk for All Communities*, Washington, DC: Government Printing Office.

—— (1995) *Waste Programs Environmental Justice Accomplishments Report – Factsheet*, Washington, DC: USEPA Office of Solid Waste and Emergency Response.

United States General Accounting Office (USGAO) (1983) *Hazardous and Nonhazardous Waste: Demographics of People Living near Waste Facilities*, Washington, DC: Government Printing Office.

Valetta, W. (1993) 'Siting public facilities on a fair share basis in New York City', *The Urban Lawyer*, 25 (1): 1–20.

Villiers, B. de, van Vuuren, D.J. and Wiechers, M. (1992) *Human Rights, Documents that Paved the Way*, Pretoria: Human Sciences Research Council.

Vogel, L. (1994) *The Fragile 'We', Ethical Implications of Heidegger's Being and Time*, Evanston, Ill.: Northwestern University Press.

Walker, D. (1995a) 'Eastern sale to fetch $2 billion', *The Age*, Melbourne, 6 November, p. 7.

—— (1995b) 'Victorians should use more electricity, says US buyer', *The Age*, Melbourne, 7 November, p. 3.

Walker, M. (1993) 'Global taxation: paying for peace', *World Policy Journal*, Summer.

Walker, R. (1981) 'A theory of suburbanization: capitalism and the construction of urban space in the United States', in Dear, M. and Scott, A.J. (eds) *Urbanization and Urban Planning in Capitalist Societies*, London: Methuen, pp. 383–430.

Wall, J. A. (1981) 'Mediation: an analysis, review, and proposed research', *Journal of Conflict Resolution*, 25: 157–89.

Walzer, M. (1983) *Spheres of Justice, A Defence of Pluralism and Equality*, New York: Basic Books.

Waring, M. (1988) *If Women Counted, A New Feminist Economics*, London: Macmillan.

Warren, K. (1990) 'The power and the promise of ecological feminism', *Environmental Ethics*, 12/2: 125–46.

Watkins, K. (1996) 'Loss of individual rights as WTO charts course for global free trade', Canberra, Australia: *Canberra Times*, 8 December, p. 5

Weale, A. (1992) *The New Politics of Pollution*, Manchester: Manchester University Press.

Weber, M. (1964) *Wirtschaft und Gesellschaft, Studienausgabe*, based on the 4th German edn, 2 vols, Johannes Winckelmann (ed.), Köln: Kiepenheurer und Witsch.

Weir, D. (1987) *The Bhopal Syndrome: Pesticides, Environment and Health*, San Francisco: Sierra Club.

Weizsäcker, E. von (1994) *Earth Politics*, London and New Jersey: Zed Books.

Wenner, L. (1993) 'Transboundary problems in international law', in Kamieniecki, S. (ed.) *Environmental Politics in the International Arena, Movements, Parties, Organizations and Policy*, Albany: State University of New York Press.

Wenz, P.S. (1988) *Environmental Justice*, New York: State University of New York Press.

Weston, A. (1985) 'Beyond intrinsic value: pragmatism in environmental ethics', *Environmental Ethics*, 7/4: 321–40.

Weston, J. (1986) 'The Greens, "Nature" and the social environment', in Weston, J. (ed.) *Red and Green: The New Politics of the Environment*, London: Pluto Press, 1–29.

Wolf, Z. (1992) 'The massive degradation of ecosystems in the USSR', in Stewart, J.M. (ed.) *The Soviet Environment: Problems, Policies and Politics*, Cambridge: Cambridge University Press, pp. 40–56.

Wood, A. (1981) 'Marx and equality', in Mepham, J. and Ruben, D. (eds) *Issues in Marxist Philosophy*, vol. 4, Brighton: Harvester Press, 195–221.

Young, I.M. (1990) *Justice and the Politics of Difference*, Princeton, NJ: Princeton University Press.

Young, J.E. and Sachs, A. (1995) 'Creating a sustainable materials economy', in Brown, L. (ed.) *State of the World 1995*, London: Earthscan Publications, pp. 76–94.

Young, O.R. (1994) *International Governance, Protecting the Environment in a Stateless Society*, Ithaca, NY and London: Cornell University Press.

Young, R. (1992) 'Egalitarianism and personal desert', *Ethics*, 102: 319–41.

Zimmerman, M.E. (1993a) 'Heidegger, Buddhism and deep ecology', in Guignon, C.B. (ed.) *The Cambridge Companion to Heidegger*, Cambridge: Cambridge University Press, pp. 240–69.

—— (1993b) 'Rethinking the Heidegger–deep ecology relationship', *Environmental Ethics*, 15: 195–224.

—— (1994) *Contesting Earth's Future, Radical Ecology and Postmodernity*, Berkeley: University of California Press.

NAME INDEX

Abbey, E. 144
Ackroyd, P. 80–1
Adeola, F. O. 108
Allende, I. 29
Alston, D. 107
Altvater, E. 61, 64, 129, 170, 171–2, 173–4
Anderson, P. W. 122
Anderson, T. 162
Anderson, W. 125
Apel, K.-O. 24, 72, 94–5
Archibugi, D. 185, 207
Archimedes 97
Arendt, H. 22, 47–8
Aristotle 51, 72
Armour, A. 116
Attfield, R. 142

Badcock, B. 106
Bahro, R. 169
Baier, A. 89
Baker, R. 10
Barry, B. 85
Barry, J. 59
Barthes, R. 37
Bateman, I. J. 162
Bateson, G. 150, 153
Bauman, Z. 36–7, 38, 45, 51, 99
Beck, U. 5, 14, 16, 17, 25, 60, 64, 76, 103, 104,122, 126, 127, 131–2, 167, 168, 174, 180
Beckerman, W. 161, 162, 163, 164
Bell, D. 135, 136
Bellah, R. 40, 91
Benhabib, S. 22, 38, 47–8, 92–3, 200
Bennett, J. 162
Bentham, J. 73, 75, 77, 84, 133, 135

Benton, T. 61–2, 64, 140, 141, 143, 155, 174
Berger, P. L. 34
Bhaskar, R. 20, 53, 195, 196, 197
Birnie, P. 178
Blair, T. 198
Block, W. 162
Blomley, N. K. 105
Blowers, A. 122, 128–9, 130, 164–5, 166, 167, 168
Boerner, C. 116
Bogard, W. 126
Bookchin, M. 36, 39, 59, 113, 143–5, 145–6, 168, 175
Bottomore, T. 33
Boyne, G. A. 105
Brechin, S. R. 58
Brennan, A. 68
Brian, A. W. 122
Bromley, D. W. 161
Brooks, R. O. 112
Brubaker, E. 161
Bryant, B. 108
Bullard, R. 108, 117, 118
Burnheim, J. 188, 190, 191, 192
Burton, J. 95

Calicott, J. B. 139
Capra, F. 2, 149
Carson, R. 136
Carter, F. W. 11
Carver, S. 115
Cassese, A. 177
Castells, M. 106
Chan, Sir J. 10
Charnovitz, S. 121
Charvet, J. 92
Childers, E. 181

SUBJECT INDEX

SUBJECT INDEX

of 84; industrial 122, 167; international, necessary feature of 170; international, uniform constitutional order 188; liberalism of societies 40; limited by external factors 172; mode of production 169; modern, constitutional order imposed upon colonised nations 187; 'pale green' 210; political virtue 189; predatory and invasive 84; regulating 174, 178; restraining 190; self-valorisation 172; short-term interests 67; social contradiction of 64; sustainability of 171; technology and 148; tendency to distribute socio-economic and environmental resources unevenly 106; transition to 11; uneven development on which it thrives 183; unrestrained, unqualified defence of 79; welfare 64; see also global capitalism

care 22, 44–5, 57, 67, 89, 91; balance between freedom and 204; mutual 135; rights to 58

'catallaxy' 80, 82

catastrophes: ecological 126; global environmental 168

categorical imperative 56, 93, 151, 201

Central America 128

cereals 124

chaos 99

chemicals: hazardous 108, 212; poisoning 114, 127; toxic 124

Chernobyl 11

children 108

China 10, 181

choice(s) 46, 179; freedom of 51, 79; moral 38; political 190; protection of 60; public 74, 160, 161, 164; realising 43; social 33–6; wider horizons of self 200–1

Christianity 87, 175–6

churches 107

cities 148, 175, 176; capitalist 106; housing and amenity 122; industrial 102; major 105

citizenship 57, 58, 113, 139, 201; 'good' 59; ideals of 60

civil disobedience 108

civil rights 58, 108; organisations 107

class 62, 103, 108, 113, 186; conflict 32; environmentally privileged 116;

exploitative structure 37; inequality 58, 106; middle 107, 118, 122; population differentiated by 208; structured society 29; struggle 66; varying access to knowledge and information 118; working 66, 122, 182

climate change 2, 180, 181; anthropogenic 206

climatologists 210

coal 35, 181

coercion 185; see also uncoerced discourse

cognitive abilities 47

collective action 80–1

Colombia (Bogota) 54

colonialism 55, 82, 177, 187, 188

colour see people of colour

commonwealth 56, 58

communication 61, 68, 93, 94; human and non-human nature 97; truth aspired to and sought by 197

communism 33, 64; imminent collapse of 196

communitarianism 23, 73, 90–2, 99, 138–9; critique of justice 29; critique of liberalism 39–42; ideas put to the test 204; relativism encouraged by 118

communities 175; biotic 39, 138, 139, 147; bounded 41; coloured 116, 117; 'community of' 41, 92; 'consensual' 118; ecological 39; empowerment of 113; environmental costs must be shifted from 205; feudal 58; global 24, 41, 42, 204; 'good' and 'bad' land uses between individuals and 103; heightened environmental risk for 130; homogeneous 190; identification with 134; imaginary 187; indigenous 118; land 99; life-forms and 157, 200; 'loser' 116; moral 65; native, pressure by waste corporations on 118; non-human and human, equity between 148; organic 40; political 58; poor 116, 117, 119; residential 114; risk to 123; self-sufficient 39, 42; universal 41; wealthier 118; see also local communities; minority communities

compassion 57, 58, 60, 68, 69

compensation 8–9, 50, 52, 53, 98, 110,

Indexes compiled by Frank Pert